Growing
ORCHIDS

EXPANDING YOUR ORCHID COLLECTION

By the same author

Growing Orchids:
Book One, Cymbidiums and Slippers
Book Two, The Cattleyas and Other Epiphytes
Book Three, Vandas, Dendrobiums and Others
Book Four, The Australasian Families
The Specialist Orchid Grower

Growing
ORCHIDS

EXPANDING YOUR ORCHID C·O·L·L·E·C·T·I·O·N

J.N. RENTOUL

Lothian Publishing Company Pty Ltd
Melbourne · Sydney · Auckland

A Lothian Book
Lothian Publishing Company Pty Ltd
11 Munro Street, Port Melbourne, Victoria 3207

First published 1989

National Library of Australia
Cataloguing in publication data:

Rentoul, J. N. (James Noel), 1909-
Growing orchids. Expanding you orchid collection.

ISBN 0 85091 364 0.

1. Orchid culture. I. Title. II. Title:
Expanding your orchid collection.

635.9'415

Typeset in Garamond by Best-set Typesetter Ltd
Printed in Singapore by Kyodo Shing Loong Printing Industries Pte Ltd

Front cover picture: *Laelia purpurata*

Contents and List of Genera

Introduction	*vii*	*Dendrobium*	54
Acampe	1	*Dendrochilum*	59
Acineta	2	*Disa*	60
Ada	3	*Doritis*	63
Aerangis	4	*Encyclia*	64
Aeranthes	5	*Epidendrum*	67
Aerides	6	*Epigenium*	69
Amesiella	9	*Eria*	70
Ancistrochilus	10	*Eulophia*	71
Angraecum	11	*Galeandra*	73
Anguloa	14	*Gongora*	75
Ansellia	16	*Grammangis*	77
Arachnis	17	*Grammatophyllum*	78
Arpophyllum	19	*Grobya*	79
Arundina	20	*Habenaria*	80
Ascocentrum	21	*Haraella*	82
Aspasia	22	*Hexisia*	83
Baptistonia	23	*Huntleya*	84
Barkeria	24	*Laelia*	85
Bifrenaria	25	*Lemboglossum*	90
Brassavola	27	*Leptotes*	92
Brassia	29	*Ludisia*	93
Broughtonia	31	*Lycaste*	94
Bulbophyllum	32	*Masdevallia*	95
Calanthe	34	*Maxillaria*	98
Catasetum	36	The Plates	101
Cattleya	39	*Meiracyllium*	205
Ceratostylis	44	*Miltonia*	206
Chiloschista	45	*Miltonioides*	207
Chondrorhynca	46	*Miltoniopsis*	208
Chysis	36	*Mormodes*	36
Cirrhaea	47	*Nageliella*	210
Cochleanthes	46	*Neobenthamia*	211
Comparettia	48	*Neofinetia*	212
Constantia	49	*Neolehmannia*	213
Cryptochilus	50	*Neomoorea*	214
Cuitlauzina	51	*Notylia*	215
Cycnoches	36	*Oberonia*	216
Cynorkis	52	*Odontoglossum*	217
Cypripedium	53	*Oncidium*	219

Ornithocephalus	222
Pabstia	223
Palumbina	224
Panisea	225
Paphiopedilum	226
Peristeria	228
Phalaenopsis	229
Phragmipedium	232
Phreatia	233
Pleione	234
Pleurothallis	235
Polystachya	236
Promenaea	237
Renanthera	238
Restrepia	239
Rhyncolaelia	240
Rhyncostylis	241
Robiquetia	242
Rodrigueziella	243
Rossioglossum	244
Saccolabium	241
Scuticaria	245
Sigmatostalix	246
Sobralia	247
Sophronitella	248
Sophronitis	249
Spathoglottis	251
Spiranthes	252
Stanhopea	253
Stelis	254
Stenocoryne	255
Symphyglossum	256
Thunia	257
Trichoglottis	258
Trichopilia	259
Vanda	260
Warscewiczella	46
Vandopsis	263
Zygopetalum	264
Zygostates	265
Bibliography	267
Index	271

Introduction

In the 200-odd years in which orchids have been grown as a pastime and commercially, botanists, naturalists, horticulturists and ordinary cultivators have tried to assess the number of species in the world's orchid flora. It is like trying to count the stars in the sky, and about as useful. The number has been put at about 25,000 to 30,000 species, belonging to some 1200 to 1300 genera. Orchids are one of the largest families of plants.

One has to ask also, how much of the orchid flora is left after those 200 years? Some years ago in *Growing Orchids, Book Four* I worked out an answer for the Australian content of the world's total; I estimated that after 200 years' occupation of the continent, only 2 per cent of the original number of orchid plants remains. That could be an optimistic figure. A major cause of loss of plants is the destruction of habitats which is still unfortunately proceeding.

In its original form the cult was based on collection of species orchids of all available genera; these were transferred to artificial conditions which subsequently killed them almost as fast as they became available. As knowledge increased there was an improvement and the records show magnificent specimens were grown by various people. But in general these specimens were not produced artificially — they were frequently the result of plant disturbance provoking flowering which was very hard to duplicate with the same material. That the material was so easy to replace was most regrettable as it entailed the almost complete destruction of the source plants.

Following in the wake of the collectors were those who converted the land to food production, usually by the easiest method of deforestation and burning the refuse. In most instances this also included the plants — not only orchids but the ferns, bromeliads, the mosses and lichens which make up the ecological background in which orchid plants survive and multiply.

The cultivation of species orchids has persisted right through the pastime from its inception, sometimes by deliberate choice and excluding hybrids, and at other times as part of a complete collection. In the middle years about the end of the 19th century, production of hybrid orchids brought new ideas and new flowers. There was sufficient impetus in the innovative ideas about hybridising to give the species orchids in all countries a breathing space. But the day of the collector and gatherer came to an end.

Orchids did not become widely popular until the cymbidium was produced in endless numbers by new seed-raising techniques. The fashions in orchid growing changed: by the 1930s the cymbidium had almost taken over and those who grew species orchids became the minority. The plants which once drew notice no longer commanded attention. This was brought home to me in the 1940s when I produced a plant of *Masdevallia coccinea* (Lindley), one of the best and largest forms, with fourteen beautiful flowers on stems 45 centimetres

(18 inches) high. When it was benched with the cymbidiums in the late months of the year it did not rate a mention. That hurt!

After about forty years of unchallenged popularity for its easy growing capabilities, the cymbidium has again been challenged. The once-despised species orchids are in demand over their entire range.

This trend represents a general movement in orchid culture with periodic favouritism for certain forms which are then usurped by another genus. Those following this path included almost the whole of mass-produced stock of various genera. One could well ask where all those plants named in the six volumes of *Sander's List of Orchid Hybrids* have gone. If they were worth naming, how have they disappeared so completely? No one has hazarded a guess at the number of names contained in those books, but volume one alone has 15,000-odd registrations named to the year 1945.

For the grower of the late years of the 20th century who wishes to grow species orchids in quantity the choice is still there in small numbers. To understand how the position has deteriorated, we need only consider the first thirty years of this century, when thousands of plants were still available. Then one could buy a single species paphiopedilum in lots of 100, and the plants were of four to six growths, whereas now only single growths of this genus can be obtained.

In the hundred or so genera reviewed in this book, several factors are brought to the notice of growers. It could be thought that at times the requirements are rigidly laid down, but they are better regarded as primary rules which can be bent a little. This flexibility, however, should not be mistaken. In nearly all instances temperatures may go a little higher, but never lower. Information given about humidity and shading is based on a particular climate: equable, slightly sub-temperate, cool, but never with freezing temperatures in winter.

Considerable thought must be given to the cultivation of a loosely related collection of plants which have come from several different climates and environments. Some compromises must be made so that each species adjusts its preferences upward or downward a little to make them fit a pattern. Something is lost from each, but with good attention to details like micro-climate and correct benching they can be fitted together so that none are lost.

There should always be an understanding on buying species orchids that even in nature they are never meant to associate. Each has a separate environment. In the instance of paphiopedilums, of which there are between fifty and sixty species in the genus which are readily recognisable as separate entities rather than variants of a known species, few of the plants could be placed in natural surroundings and tried as subjects for re-establishment one beside another. Left to their own devices, if any at all survived, the number would be drastically reduced by the end of the second year and few indeed left to survive the fourth annual cycle of growth and flowering.

It is rarely possible to grow species orchids artificially and then transplant them on establishment to reconstitute a habitat. I regard that as the tertiary phase in the life of any orchid, terrestrial, epiphyte or lithophyte. The rule is that orchids grow where orchids grow. They seldom die out completely if left to their own devices in their natural habitat, however harsh that may appear to be.

In effect, we are looking at the primary rule — understand the limitations of what you, as a grower, have to offer the plants. Understand, too, the nature of the plant: it may be quite different in its needs from others you already have. For example, if you have *Cymbidium lowianum* (Reichenbach f.), one of the easiest of the genus to grow and flower, it would not follow that you could buy a plant of *Cymbidium finlaysonianum* (Paxton) and grow it in the same conditions alongside it. One comes from the Himalayan foothills region, the other from the Borneo–Sumatra region.

Similarly with all the genera outlined in this book. Some masdevallias — of which there are between 300 and 400, and possibly more yet to be named — grow in cool artificial conditions, others in temperate to moderate warmth and a few need a little more than that to grow successfully and produce flowers.

The matter of temperature is the most important. Where plants are recommended for cool conditions that does not mean cold. Too frequently the terms 'cool' and 'warm' are regarded as relative to the climate in which one lives. In a tropical area, for instance, those used to a climate where the average daily temperature is 30 degrees (85°F.) will regard a drop to 20 degrees (68°F.) as cool. Yet that lower temperature is a good minimum for warm-growing orchids day or night over the whole year. That is where the flexibility of upward movement is best illustrated.

The table of temperatures does not convey absolute limits; it is a guide to those unsure about the limitations of their growing facility and does not express anything unusual such as requirements for tropical cultivation. These are given in considering the various genera for that type of climate.

It should also be impressed here that species orchids should preferably be chosen to suit the environment in which each grower lives. If that is ignored then heating or cooling devices should be considered. These add to the cost of the pastime, but it is better to make them part of the program rather than to put plants at risk by growing them in unsuitable conditions.

RECOMMENDED TEMPERATURES FOR ORCHIDS (DEGREES CELSIUS)

	Minimum (daily)	*Maximum (daily)*
Cool	10 (50°F.)	27 (80°F.)
Intermediate	15 (60°F.)	36 (100°F.)
Warm	18 (65°F.)	36 (100°F.)

The table lays down arbitrary divisions only. It is almost impossible to adhere strictly to the figures and a certain amount of fluctuation is expected. It will be noted that there is an upper limit beyond which most orchids are at risk unless controlling influences are brought into play, such as extra shading, increased ventilation and application of surface water to the surroundings but not the plants. It is very poor cultural technique to try lowering the temperature by spraying the foliage of orchids in these circumstances.

Some species in the cool section will not readily tolerate the upper limit and are poor mixers with orchids which are tolerant. Such a species is *Odontoglossum crispum* (Lindley) which, unlike the hybrid odontiodas and odontoglossums, will not tolerate temperatures above 20 degrees (68°F.).

The genera's requirements for shade are given in percentages according to which shade cloth is commonly sold. Knitted cloths have slightly more sun-breaking power than woven cloths. In relation to shade, it is best to give the brightest factor the plants will stand, rather than grade the shade down to suit the lowest common factor.

Humidity, one of the most important things in plant growth for orchids, may be gauged accurately by a fairly cheap meter. In high temperatures the humidity will fall unless there is residual moisture available for evaporation. The highest measure of humidity is not always obtained in the high temperatures, however: in cool conditions it may move into undesirably high figures. In such instances it should be dissipated by admitting more air, however cool that air may be. The ideal range, if it can be stabilised, is between 40 and 60 per cent for the cool and intermediate grades and from 60 per cent upward for the warm-growing orchids.

The brief review of the more than a hundred genera in the following pages gives enough material to keep the average orchid cultivator busy for a lifetime. It should be apparent to growers that the prevailing climatic pattern of the area where they live is the criterion by which to judge what shall be bought or accepted as a gift. Perhaps that is the most difficult lesson to be learned in this absorbing pastime. It is better to refuse than risk any species orchid of whatever genus in a situation in which there is a realistic chance that it will die.

In a genus there may be hundreds of species to choose from, or perhaps only one. The chances of obtaining a particular species may be remote and it may be necessary to exchange or to buy seedlings. There is value in settling for only the best, seeking not the largest plants, but rarity and beauty. It is wise also to exercise foresight and selective capacity to choose plants which will become popular.

Propagation of worthwhile clones should be the aim of every species grower. If you are pollinating in order to collect seed, always try for cross-pollination rather than self-pollination. In this way the integrity of the genus is assured.

The last piece of advice, and perhaps the most important, is not to overcrowd the collection. Small plants will always grow into large plants if that is their nature. Assess their potential as you start and allow for the increase.

Not all the cultivated genera are described here, but there is enough material to give scope for all three temperature ranges. The most important of these, the middle or intermediate section, is the best to aim for, and it has a large amount of latitude.

The Genera

Acampe (Lindley) 1853

A genus of principally summer-flowering vandaceous orchids with an extended occurrence reaching from tropical Africa to India and Burma. They first came into British cultivation with the importation of *Acampe multiflora* (Lindley) from northern India very early in the nineteenth century. It was flowered in William Cattley's collection, probably under the synonym *Vanda multiflora* (Lindley) and was featured in Lindley's *Collectanea Botanica* as illustration number 38.

There are few species in the genus and most could be termed intermediate climate orchids. This section is generally regarded as capable of cultivation in minimum temperatures of preferably not less than 15 degrees (60°F.). This range of temperatures is quoted throughout the book and it represents the grade of plants which exist between the cool-growing genera and those better named tropical. The former include some laelias, and the latter is well represented by phalaenopsis.

Vandaceous orchids generally are ill-suited to pot culture but an alternative is difficult to devise. The best alternative to large pots is the slat basket, which, of course, must be suspended. In some ways this suits the vandaceous group better than others because it gives the trailing root systems scope to move into the air below the containers. In nature, the trailing root systems are not apparent: instead, the roots prefer to travel along tree trunks or branches or other surfaces. It is not until plants become pendant that the roots frequently remain aerial.

The genus acampe mostly has plants of stout morphology, with thick leaves which are channelled and shortish flower stems on which the blooms are frequently crowded, as on page 101.

Most of the genus are light-hungry rather than shade-loving, but this must be tempered in cultivation, particularly under glass, so that the foliage is not burnt. Shade cloth of about 75 per cent is needed in temperate to warm climates where bright sunlight occurs through warmer months of the year.

Humidity in this period is important to most vandaceous plants. If coupled with correct light it should rate at anything from 60 to 70 per cent but should not reach the 100 per cent liked by phalaenopsis and other tropical genera.

Fertilisers for acampes can be liquid or solid. If the roots can be trained entirely into containers, either is appropriate. But if most of the root is aerial the fertiliser should be applied as a spray. Flowering is variable, spring to autumn.

Acineta (Lindley) 1843

A small genus of epiphytic orchids native to Central America, Colombia and Venezuela. It is related indirectly to stanhopeas, but to very few other genera of the region. The habit of the plants is like stanhopeas, but the foliage is broader and longer, something like that of peristerias. The flower spikes form from the base of pseudo-bulbs and are pendant.

The genus is principally tropical in nature, growing naturally in fairly high temperatures in some parts of the annual cycle. As cultivated orchids they fall into the category between intermediate and warm subjects, with emphasis on temperatures of about 15 degrees (60°F.) and upward to tolerable levels.

The foliage indicates the degree of protection needed. It is broad and could be termed insubstantial compared with that of most stanhopeas. The foliage is all concentrated on the apex of pseudo-bulbs and usually comprises three or four leaves, those inside the whorl being shorter than the outer foliage. The plants are best suited to slat basket culture.

The Mexican species *Acineta barkeri* (Lindley) was once known as *Peristeria barkeri*. It was beautifully reproduced in Bateman's *Orchidaceae of Mexico and Guatemala* from a painting by Mrs Withers in water-colour. This artist was flower painter to Queen Adelaide (1792–1849), wife of King William IV.

The collector Warscewicz found *Acineta densa* (Lindley) in 1819 and gave it to G. Ure Skinner. He sent it to the Horticultural Society. It was noted that the flowering racemes grew more than 60 centimetres (2 feet) long. By comparison the illustration, growing on a small plant with a few flowers, is poor. *A. densa* (Lindley) is correctly known as *Acineta chrysantha* (Morren).

The climate needed to cultivate these orchids is mildly tropical, but it should not be heavily humid and extreme. This type of climate may be produced by air injection through a fan system and brought further into line by correct shading of about 60 to 75 per cent. In climates which alternate between hot and dry and cool and moist a steady temperature range between 15 and 25 degrees (60–77°F.) would be an objective. In such conditions the alternating climate itself would induce a growing, flowering and dormant pattern.

Such patterns dictate watering and fertilising, very necessary functions for plants such as acinetas that produce long spikes of flowers. Solid fertilisers should be avoided and liquid nutrients relied on to maintain plant health. Complete dehydration during dormancy should be avoided.

Ada (Lindley) 1853

A genus of four members, two of which were included during the 1970s and 1980s. It was once considered to have a single member in *A. aurantiaca* (Lindley); a second species, *A. lehmannii* (Rolfe) was added, named in honour of a one-time consul for Germany in the state of Colombia, a South American country on the north-western flank of that continent, known in early years as New Granada. Lehmann died in unfortunate circumstances, as related in *Growing Orchids, Book Three.*

The third species was added in 1972. Originally known as a brassia, *Ada allenii* (L. O. Williams ex Schweinfurth) N. H. Williams, more closely relates to ada than to its original genus. See page 102.

The Venezuelan *Brassia glumacea* (Lindley) was the fourth inclusion, but is probably better known by its older title.

Schlim collected the first plants of the genus and sent them to England and Europe. James Bateman, who detested hybrids and hybridising of species, flowered the first plants about 1851. The history is more fully given in *Growing Orchids, Book Three.*

Morphologically the first two species are more distinct. *Ada aurantiaca* is the taller with foliage about 30 centimetres (12 inches) high. *A. lehmannii* has lightly marbled foliage, which is slightly shorter, and the flower has a white labellum. The pseudo-bulbs of each are slender and soft-looking and compressed. The flower spikes appear inside the longest of the sheathing leaves in spring.

A. allenii, which is illustrated, is golden yellow, the flowers open more fully and the flower spikes are erect. The pseudo-bulbs, unlike those of *A. aurantiaca*, may flower more than once. The leaves are more fan-like and the enclosing bracts of early growth about the pseudo-bulbs are untidy-looking and obvious. It is Central American.

Cultivation of all is similar to that for odontoglossums, but while *A. aurantiaca* and *A. lehmannii* should never be allowed to dry off, the other two species have more resistance to slight aridity. All originate at high altitudes in cool forest, with the Colombian species discovered at about 2500 metres (8000 feet). *Ada aurantiaca* is also found in western Venezuela.

Pot culture suits all adas, although in nature they were occasionally lithophytes. The root systems are coarse, slightly more so than odontoglossums. Potting material should not be too fine, and drainage should be good. The plants require temperatures ranging from moderately cool to intermediate, and shade in the 75 per cent range. Liquid fertilisers during growth and flowering benefit the plants.

It is possible other species may be transferred to the genus.

Aerangis (Reichenbach f.) 1865

The orchid grower cannot help but wonder why the angraecoid group has been broken up into so many sections and allocated names by successions of botanists and others. In addition to some closely related species, the study of some of these variations brings into focus the curious resemblance of the two Philippine and Japanese orchids amesiella and neofinetia to the angraecoid orchids in their entirety. All have a vandaceous resemblance. No doubt many are puzzled by the distribution of these orchids in the Afro-Asian sphere.

Aerangis as a genus is distinct from rangaeris and ysilopis, although all are fundamentally similar. The distinction between rangaeris and aerangis and the similar species ysilopis, as given by Brian Morris in *Orchids of Malawi*, is a 'wide variance in certain microscopic characters — principally the structure of the column and pollinarium'. It is doubtful if an orchid grower other than one intent on these botanical variations would be able to recognise the orchids by their morphology, whether of plants or flowers.

Unfortunately African orchids of any genus are difficult to obtain. They are principally epiphytic, hardy and need no more than basic adherence to intermediate cultural levels.

Aerangis are fairly common orchids in various parts of southern and middle Africa. Some show preference for rather bright light and do not seem to be selective about the host trees. They have been known to grow on 'exotic cypress trees' according to Brian Morris. The usual habitats are fairly well shaded but in moderately dry locations with definite seasons.

In cultivating the plants most of these smaller species should thrive in intermediate climates under about 50 to 75 per cent shade. The leaves are usually bilobed at the apices, some markedly so, and grey-green to deeper green. They vary in length, according to the species. There are usually four to six leaves and they are impermanent, unlike those of other vandaceous types. The number of flowers varies with differences in culture and the nature of each species, some plants bearing more than one spike of flowers and at times up to four. The usual spur associated with the complex of genera is always present and of varying lengths.

There is a definite resting period in all the species, indicated by the root tips. The roots grow longer but branch only in few instances.

Aeranthes (Lindley) 1824

The genus comprises about 30 species, many of which were known at various stages of their history as angraecums or aerangis. Most are Madagascan orchids, monopodials and usually with sparse leafage. The climate in which many grow is rather harsh compared with the softer types encountered in semi-tropical and tropical areas. The genus is found almost throughout the African region.

As outlined in the section devoted to angraecums in this book and *Growing Orchids, Book Three*, the interrelationship of the group is based principally on recognisable characteristics in the flowers such as the long spur extending from the base of the column and labellum. In some species of the group this spur is degenerate and the flowers themselves are frequently cast before maturity or opening, with a single specimen remaining. This could be noted in the species *Aeranthes africana* (Stewart), which was a late discovery of the year 1974 in Rhodesia, growing at more than 1600 metres (more than 5000 feet). John Ball records it in his book *Southern African Epiphytic Orchids* as the sole representative of the genus in mainland Africa.

When considering addition of the angraecum-like section to an existing collection, it is as well to understand the nature of the plants and the climate required to keep them growing and flowering. They are not completely singular and should mix well with other epiphytes such as dendrobiums of the 'harder' kind. The use of the word 'harder' needs some explanation. As with many genera, the dendrobiums include plants which are regarded as tropical and best accommodated in constantly humid conditions gauged by instruments as varying between 50 and 75 per cent. Others, which include those 'harder' species, frequently experience periods of much lower humidity which may last for some months. *Dendrobium bigibbum* falls into this category, although the change in its growing period is dramatic and may vary up into quite high limits in the 80 to 100 per cent humidity range for some months and in bright light.

Aeranthes need this kind of artificial climate. The rules are not hard and fast but variation into lower ranges of temperature and high humidity would be completely unsuitable.

Growing methods for aeranthes should preferably be suspension rather than benching. If the plants are potted, the material should be coarse and the plants perhaps mounted on treefern poles inserted into the pots so that root growth or dormancy may be noted. Watering and fertilising would be indicated by the growing tips of the roots. The flowers, particularly the spurs, should be kept free of moisture.

Aerides (Loureiro) 1790

This genus is very widespread, reaching from India and Sri Lanka in the west through Asia to the Philippines and southern islands of the West Pacific.

Loureiro named the genus in the belief that these plants obtained all their nourishment from the air since they were not terrestrial. The moistness of the environment, of course, has much to do with their cultivation and health. Loureiro was a Portuguese priest of the eighteenth century who was more concerned with the various plants, including orchids, of Cochin China, today known as Vietnam.

The plants grow in many climates, never totally isolated to set and rigid temperatures, but preferring in general climates between the temperatures of 15 and 30 degrees (60–85°F.). Some species are listed as cool-growing, among them *Aerides multiflora* (Roxburgh) and *A. vandarum* (Reichenbach f.). Even these species prefer slightly warmer than cold conditions, despite the fact that they are frequently regarded as cold-growing aerides.

The genus is monopodial — that is, nearly all the plants grow as single stems which may or may not branch as they become taller and stronger. The extent to which this characteristic occurs indicates good culture, in which the leaves remain on the single stem of the plant even if it branches. In cultivation it is possible to partly cut through the main stem to accentuate the branching habit, as described in *Growing Orchids, Book Three*.

The root system of vandaceous orchids such as aerides issues from the main stem of the plant and there is no value to be placed on the points at which a root may bud from the stem. They appear probably at some point which the plant 'thinks' will lead to a food source. Some of the ways in which I have sought to influence this have proved nothing. But it certainly has a lot to do with humidity and air flow.

All vandaceous orchids are open-rooted plants. Although they may be grown in containers, it is most probable that the vigorous root systems which good culture produces will be principally open to the air, and efforts to contain them are against the nature of the plants and will be resisted.

It would appear from that statement that fertilising is a problem, but the roots will seek out sources of the nutrients they need. Fertilisers may be incorporated in potting mixes as solids in the form of organic material, or applied as inorganic liquids which soak into the potting mix.

Leaves may vary in species from the 'strap' type — almost flat leaves with a slight centre channel — to the terete type — rounded and of various lengths, somewhat like short coarse knitting needles. The latter type of leaf indicates that the plants originate in bright light, and this information will dictate the environment in which they are grown. *Aerides vandarum* (Reichenbach f.) is representative of the type.

Most members of the genus are best suited when attached to some form of support such as a section of treefern 5 or 6 centimetres square and about 60 or more centimetres long (2 inches square by 24–30 inches tall). This may be held in a pot or container filled with stone aggregate for firmness and pot stability. Alternatively, use a pole of the same size made of weathered non-resinous wood.

Plants may be tied to such poles with nylon fishing line or strips of nylon from discarded stockings or other durable material. Once the roots adhere to the wood or enter the treefern pole, part of the cultural problem is solved. If they do not attach, discard the pole and try another. Some treefern poles are markedly acid and discourage roots.

Fertilising should be carried out by watering the top of the pole rather than spraying over the plant itself. The nutrient liquid or water will saturate the material and in this way the root system will be able to find all the fertiliser it needs.

The needs of monopodial aerides are moderate rather than avid almost throughout the growing and flowering phases of culture, and it will be noted that a healthy penetrating root system will aid flowering.

The natural habits of aerides are to proliferate from branches sent out from the main stem; if these grow they broaden the scope of the plant and add to the flowering potential of some aerides. *Aerides vandarum* mentioned above usually has only two or three flowers on each raceme, but if there are several growths the effect is far different.

Strap-leaf aerides are best attached to poles if they are inclined to grow tall. Some may be slow-growing, and others again grow and flower better if left to their pendant habits. In some instances the root systems may be perverse and adopt straggling habits, making the feeding of nutrients a problem. In these instances the roots may be attracted to fertilised material in a container. It should always be coarse and absorptive rather than solid.

In growing species orchids, the problems of supply make some items a rarity. Unless some source is located it is usually a long wait until plants are obtainable from dispersal sales. Any orchid plant's price is directly related to supply and demand and finally what it is worth as an addition to a collection. There is a great amount of similarity in aerides species, although some diminutives are available in allied genera which are worth obtaining. Any species orchid should be judged solely as an orchid and not always for its size.

A short list of aerides, some of which need more than usual cultivation care, is:

A. *crassifolia* (Parish and Reichenbach f.), Burma. Intermediate conditions, recommended minimum temperature of 12 degrees (54°F.) at any time.

A. *crispa* (Lindley), India. Intermediate.

A. *lawrenceae* (Reichenbach f.), Philippine islands. Intermediate.

A. *multiflora* (Roxburgh), India and Burma. Intermediate.

A. *odorata* (Loureiro), one of the most widespread of the genus, which also has some twelve synonyms, from India, Burma, to China, Thailand, Malaysia and the Philippine islands. Intermediate.

A. quinquevulnera (Lindley), Philippine islands. Intermediate to warm.

A. vandarum (Reichenbach f.), India. Cool to intermediate.

There are some 60-odd members of the genus, all of which should be considered rainforest species needing the light, air flow and moisture associated with such an environment.

In a mixed collection of species the genus may be grown as benched or suspended plants. Those suspended may be grown in wooden slat containers like one illustrated on page 122 and in these situations the plants may be expected to attach a certain number of the roots to the wooden slats while others will be completely aerial.

Occasionally aerides will enter a dormant period and the best indicator of this phase is that the roots become inactive at the tips. When you see this, stop giving fertiliser and reduce the moisture given directly to the plants. Renewed activity of root systems indicates need to slowly increase the water and apply fertiliser.

At all times the relative humidity of growing areas for aerides should be between 50 and 70 per cent, only occasionally lapsing below those figures.

Amesiella (Schlechter ex Garay) 1972

A genus of only one member and named in honour of Oakes Ames, a noted American botanist. As a name it is comparatively new when compared with the bulk of nomenclature.

The species comprising the genus was once known as *Angraecum philippinense* (Ames) but was excised from the genus and renamed because of noted floral differences. Its isolation is not unusual in the total orchid family, although generally speaking there are few in the some 1200-odd known members with that distinction.

As its name implies, this orchid belongs to the Philippine islands in the upper west Pacific Ocean region. It has been found in several locations, the habitats usually on forested slopes, as noted by Helen Valmayor in her books *Orchidiana Philippiniana*.

Most orchids have a little history attached to them, but this is infrequently available in modern books. In Quisumbing's *Philippine Orchids* a finding of *Angraecum philippinense* (Ames) is recorded and worth quoting here as follows:

> One day last December [1949] when they were climbing the slopes of Mayon volcano Mr and Mrs H. W. Graham met some road workers along the trail who were carrying a bundle of orchid plants in bud. It did not take long to acquire the orchids, which were then brought carefully to Manila. As the couple were leaving shortly for the United States they presented the budding plants to a friend, Mrs K. B. Day. In about a month the orchids bloomed, in clusters of glistening white flowers of unusual beauty. It was then that Mrs Day brought them to my office on January 31, 1950. She was excited and happy at her possession and not a little disappointed when I told her that the orchid was not a new species, although a rare one. The species is the lone representative of the genus angraecum in the Philippines. Bory described the genus, the name coming from angrek, the Malay name for all orchids of this habit. . .

The species was found and named about 1906 and remained included in the angraecum until separated and renamed in 1972.

In cultivation *Amesiella philippinense* may be pot grown. This may be the preferred way because of the nature of the root system and the dwarf, compact morphology. It may bear up to four flowers on well-grown plants, usually on a single spike, in late spring to summer. Once developed, the flowers should be kept clear of the potting material, which should be coarse and open. The plants prefer about 75 per cent shade in artificial cultivation.

Ancistrochilus (Rolfe) 1897

Amongst the orchids of the world many genera are made up of only a few species. Ancistrochilus is one of these, with only two species in its order. Some people hold that the genus has but one member and two varieties. Joyce Stewart in her book *Orchids of Africa* lists two species — *Ancistrochilus rothschildianus* (O'Brien) and *A. thompsonianus* (Reichenbach f.). Only *Ancistrochilus rothschildianus* (O'Brien) is illustrated. The flower on page 105 was most attractive and measured about 6 centimetres (a little more than 2 inches). Curtis's *Botanical Magazine* illustrated *A. thompsonianus* (Reichenbach f.), but neither the colour nor the plant does justice to the subject.

Native to West African locations, *A. thompsonianus* (Reichenbach f.) was first known through the collector Kalbreyer, who at the time was working for Veitch. That firm submitted the flower to the German botanist Reichenbach, who named it *Pachestoma thomsonianum*. Other botanists disliked the affiliation and Rolfe created the new genus. Pfitzer also expressed dissent from Reichenbach's decision and related ancistrochilus to the genus ipsea (1831).

Ancistrochilus grew in many habitats in tropical Africa, with high temperatures and high rainfall throughout the year. Kalbreyer found the plants at an altitude of 300 to 500 metres (1000 to 1500 feet), as epiphytes growing in association with mosses and lichens in shady forest trees.

The plants produce flowers in Australia in spring to early summer, coloured as in the illustration. But a variety more like that appearing in Curtis's illustrations was also as plentiful at an early period of their cultivation. Its labellum was almost pastel green.

The genus as a whole should be grown in tropical conditions with as little temperature variation below 15 degrees (60°F.) as possible and in constant humidity, both in growth or flowering. However, the flowers are soft and should not be 'cooked'.

A flat dish such as a clay saucer would be a better container than a pot. The alternative is a wooden slat basket, but it should not be suspended and forgotten, as the plants should not be dried out. The leaves are rather soft, resembling lycaste foliage, but they do not grow as tall, and are less light resistant. Shade should be constant through the year to prevent sunburn. Potting material should be coarse cymbidium type mix of bark and drainage aggregate.

Fertilisers should be applied as growth proceeds and during flower production; then stop giving fertiliser so plants can rest in slightly dryer conditions.

Angraecum (Bory) 1804

The genus is related to several other orchids of Africa and contiguous islands. They include aeranthes (Lindley) 1824, aerangis (Reichenbach f.) 1865, angraecopsis (Kranzlin) 1901, mystacidium (Lindley) 1836, rangaeris (Summerhayes) 1936, plectrelminthus (Rafinesque) 1836 and other related genera.

These orchids comprise a major African section which is almost singular to the region. Morphologically similar flowers also occur in genera such as amesiella and neofinetia in the Indo-Asian region and some similarity also extends to vandaceous orchids of the same region.

Angraecum as a genus was the source material of a book by the French naturalist and collector Louise-Marie Aubert-Aubert Du Petit-Thouars, who went to the islands to the east of Africa, including Madagascar and Mauritius, about 1790. His journey was briefly mentioned in *Growing Orchids, Book Three*. Apart from a few other reference works, his book remains almost the sole register of the distinctive *Angraecum sesquipidale* (Thouars) and the other angraecoid orchids of the island group. It has been reprinted in the original French, and its illustrations are line perfect. The taxonomists, however, over the years, have rearranged some of the nomenclature.

Bory de St Vincent, author of the generic title, was a French botanist and traveller of about the same period as Thouars, who adopted his own system of identification rather than that of Linnaeus, the Swedish naturalist and botanist who is responsible for so much of our plant classification. In all there are about 1100 orchid species in the Madagascar group of islands, some terrestrial, some epiphytic.

The orchids of Africa and its associated islands off the eastern coast, although many kilometres distant, offer the same philosophical consideration as those of Australia and its similarly associated islands, particularly if this is equated with the distribution of the genera and their origin.

The Rev. William Ellis, who first introduced *Angraecum sesquipidale* (Lindley) and *A. eburneum* (Bory) to the English fanciers, had this to say of these orchids when he returned from a visit to Madagascar:

> Both *Angraecum sesquipidale* and *A. eburneum* were the most abundant and beautiful among the orchids. I noticed that they grew most plentifully on trees with the thinnest foliage and that *A. sesquipidale* was seldom, if ever, seen on the ground, but grew high up amongst the branches.

Ellis described the root systems of these naturally growing plants as

> being seldom branched or spreading, but long, tough and single, sometimes running down the branch or trunk of a tree, between the fissures of the rough bark, to the length of 12 feet or 15 feet [3.5 to 4.5 metres], and so tough and tenacious that it required considerable force to detach or break them . . . I occasionally noticed

both *A. sesquipidale* and *A. eburneum* growing not only on the branches of living trees, but high up on the bare barked trunks of the dead trees.

It should be obvious that the habit of the plants is to be bare rooted, and such patterns in nature are the guide for artificial culture. There will be differences, of course, between the cultivation outdoors in suitable warm areas and the methods adopted in glasshouse culture. Both the species described are hardy plants thriving in almost open sunlight. In this condition the plants would assume a rather hard look, with the foliage sometimes almost shrivelling and the flowering sparse and smaller than on well-cultivated and protected specimens. However, Ellis also noted that those plants which grew in the softer material of fallen trees at ground level did not flower much compared with those exposed to bright light.

The foliage of *A. sesquipidale* (Lindley) and *A. eburneum* (Bory) is the largest and most substantial of all the plants in the angraecum group; others such as the aeranthes and aerangis have much lighter bodied foliage. Despite this, however, it is also produced to better advantage in fairly bright light and is more prone to discolouration and disease in conditions which are too humid and cold.

While many angraecums will grow quite well in containers, a lot of the root system will proceed better in the open and a small percentage only will be contained. In this they are similar to the vandaceous orchids: some form of mount should be devised to which they may be attached with nylon fishing line or strips of discarded nylon hosiery. Both are durable, and the old hosiery offers the softest and easiest procurable tie. Various mounts for attachment of the plants can be thought of, preferably well-weathered pieces of timber, long enough to give the plant ten or more years undisturbed.

Such timber mounts should be inserted into pots and weighted in position with lumps of sandstone or rock other than basalt or lava.

Treefern poles, while probably acceptable to the root systems, are not as good as timber because in less than the life of the plants they are apt to disintegrate. Cork slabs seem to vary in ways which have not been properly investigated and while the plants may attach some roots to them, the main portion of the systems will probably be aerial and unattached, particularly in glasshouses where there is humidity and good air flow.

It is possible to use ordinary containers for the larger plants such as *A. sesquipidale* (Lindley), *A. eburneum* (Bory), and *A. infundibulare* (Lindley) which is rather slender in habit and needs support. Most of the genera are self-supporting and each species shows its preference by the appearance of the plants.

I grew only two species of this genus in my orchid-cultivating career. One was *A. eburneum* (Bory), which came to me in a large seed-pan about 40 centimetres (15 inches) in diameter; it remained in that container for the years it grew in my glasshouse. The other was a much smaller delicate orchid which came with the name *Mystacidium sanderi*. The plant was small and was attached to a slice of treefern; the mount was never acceptable and it was transferred to a piece of natural cork bark, to which it attached and grew and flowered well. It had three changes of name and is now carrying the tag *Aerangis mystacidii* (Reichenbach f.) Schlechter. It is a native of the eastern sector of South Africa.

Fertilisers play an important role in cultivation of the angraecoid complex. In natural plants there appears to be some relationship between host tree and plant which has never been explained. It is still my belief, after experience with epiphytes and trees, that there is an exchange between host and what for want of a better word is classed as epiphyte. They form a close bond which can be understood only with experience and experiment.

For the angraecum genus and its associated genera, liquid fertiliser is better than solid. Apply it periodically, whenever the plants need it; this is usually judged by the state of the root system. In such open-rooted plants, of course, growing tips are the best indicators that the plants are receptive. In this state it is possible to apply fertiliser in weak and constant form whenever water appears necessary.

Natural affiliates in orchid collections suggest cattleyas, hard-cane dendrobiums, laelias; also vandas and other epiphytic orchids of intermediate temperature range, with the minimum temperature kept to no less than 15 degrees (60°F.) in the cooler months.

Anguloa (Ruiz and Pavon) 1794

There appears to be disagreement among authorities, some giving 1798 as the year of naming. The two Spanish botanists, who were accompanied in their travels to Peru by the French botanist Bombey, collected orchids in the provinces of Lima and Chancay near the coast and later inland. Their intensive research was based on Huanuco, Peru, and they also travelled and collected in Chile. Some two years following their original landing of 1778 at Callao, Peru, their whole collection made over those years was lost in a shipwreck off the Portuguese coast. They then collected a lot of new material, but much of it was lost as a result of a fire, shortly before they departed for Europe again in 1788.

Anguloa uniflora (Ruiz and Pavon) was the first of the species to be identified and lodged as the type. Colombia has more species than Peru and *Anguloa clowesii* (Lindley) was the first of its genus to flower in Britain in 1844. It is native also to Venezuela. *Anguloa ruckeri* (Lindley) (see page 106) flowered soon after and subsequently about eight or ten other species were identified and collected.

The growers Clowes and Rucker operated from about 1830 to 1850 and both reached immortality of sorts through publication of designs of their growing establishments in James Bateman's *Orchids of Mexico and Guatemala*, a beautiful publication of the brilliant years of orchid growing, printing and lithography.

Anguloas are almost totally terrestrial orchids, growing in light forest. Once that had been cleared they persisted in more or less open ground. Some plants were occasionally found as epiphytes growing in dense masses of decaying plant material. Morphologically they are very similar to lycastes, but with the base of the immature pseudo-bulbs more thickly covered with enclosing bracts. The leaves are broad and rather inclined to be soft and easily damaged in poor cultivation. They may be as long as 40 to 50 centimetres (18 to 24 inches).

Most of the species occur at about 1500 to 2500 metres (5000 to 8200 feet) in a cool to moderate climate, and this should be the mark at which to aim in artificial cultivation. The altitude may be roughly equated with distance north or south of the Equator, but the temperature should be retained between 12 and 25 degrees (54–77°F.) and plants adequately shaded considering the leaf area and its thinness.

Flowering season may be variable, but in general it occurs towards the middle of the growing process, as the flowers appear from the base of the new growth. Late spring to summer is a likely time to look for the buds and subsequent fairly constant development.

The flowers show spots if the environment is too humid, or if they are reached with the hose as they develop and open. When watering the plants, take care not to get water into the undeveloped leaf whorl as this may set up fungus rot. By the time the plants have fully developed pseudo-bulbs, the older leaves will probably have been cast, as with lycastes. But total leaf loss should be taken to indicate that the root system has been kept too wet and killed off. The plants have a rest period and should be kept rather dry until new growth and root activity commences.

Fertilise with inorganic liquid rather than slow-release pellets. A fertiliser recommended for many orchids is poultry manure pellets, but if these are used they should not be applied once the flowers are gone. Let what is already on the plant decay and disappear but do not renew the application until the growing season.

As anguloas are terrestrial type orchids a coarse cymbidium mix is suitable. Drainage should be thorough, with sufficient moisture left after watering to keep the plants damp until the next watering period.

Some shrivelling of the pseudo-bulbs is usual, much after the fashion of brassia pseudo-bulbs. This shrivelling is taken up as plants mature again after flowering, but if some of it remains it shows that they are in good health. Leaf loss is the only factor to watch. It may be regular but should never reach the stage where new leaves of the season turn yellow.

The genus as a whole is South American, with species occurring in Venezuela, Colombia, Peru and Ecuador. The distribution of the orchids in South America is such that one would expect a species common in, say, Peru to be common also in Ecuador, but a glance at a map indicates the reason why this is not so. Anguloas are found mostly on the eastern side of the mountain chain which stretches the length of South America, and in company with many other orchids live only in the climatic areas where they are suited. In cultivation these orchids should be grown in a moderate climate just below cattleya conditions, with a minimum always above 12 degrees (54°F.).

Ansellia (Lindley) 1844

A genus probably having only one member, although some may dispute this and remain convinced of the older records of several species. The best-known of these is probably *A. nilotica* (Nathaniel Brown) and *A. gigantea* (Reichenbach f.). The modern acceptance is of one species in the genus, which is solely African and from associated nearby islands.

The genus was named after the discoverer of the first specimens on the island of Fernando Po, off the west coast of mainland Africa, John Ansell. He sent the specimens of this new orchid to John Lindley about 1841, who subsequently gave its species name *Ansellia africana* in 1844.

This orchid is most variable over its range, which is considerable. In some regions it may be short, about 30 centimetres (12 inches) tall, while other varieties have been known to grow to about 1 metre (3 feet) high. In all species the dendrobium-like pseudo-bulbs are leafed mostly on the upper sections, the leaves rather thin but tolerating light which would burn most other genera. Coarse potting material suits these orchids.

Cultivation is relatively simple and similar to that of cymbidiums. The general growth period is followed by a rather dry four months, after which flowering takes place; the process repeats itself annually in the right conditions. Despite the warmth needed to grow ansellias and flower them correctly, the plants are not really tropical orchids like vandas or some of the dendrobiums they so resemble. The flowers emerge from the apex and lower on the pseudo-bulbs and are not continuous for years from the same growths. Probably ansellias would be most suited by warm-growing cattleya conditions where the plants are given less humidity at certain times of the year and moderate shade.

It is interesting to quote a writer of some sixty years ago because his summary seems to apply to the closing years of the twentieth century:

> Although containing but few representatives, the genus ansellia is worthy of attention by amateurs who desire to cultivate some of the larger-growing species and thereby include in their collections a more varied assortment of plants than is possible when hybrids only are entertained. There appears to be a return of interest in these plants, but not to the extent that was seen thirty years ago, for multitudinous hybrids now claim attention and nearly every amateur finds his houses fully occupied with seedlings that require more space each season.

Arachnis (Blume) 1825

This small genus of fewer than twenty members has habitats extending from the intermediate climates of north-eastern India through Asia to Java and parts of Irian-Jaya. The habit of most is tall to medium height monopodial growth, and they are infrequently cultivated outside tropical to sub-tropical climates.

The taxonomy of some members is confused, a typical example being *Arachnis clarkei* (Reichenbach f.), which once had the names *Arachnanthe clarkei* (Reichenbach f.) and *Esmeralda clarkei* (Rolfe), in which genus it was grouped with *Euanthe sanderiana* (Reichenbach f.) Schlechter. The confusion becomes apparent only when older literature is read. In *Reichenbachia* it is illustrated as *Arachnanthe clarkei* (Rolfe), with the inscription by Sander:

> This is a strikingly beautiful plant, with brilliant markings, introduced some seven or eight years ago, and which blossomed for the first time in this country [England] in the collection of the late Mr John Day. It is a native of the Sikkim Himalayas, and was discovered by Mr C. B. Clarke in 1875, and named by the late Professor Reichenbach in his honour.

As a genus they do not all need tropical conditions to grow and flower. Provided direct attention is given they will thrive in company with cattleyas and other orchids which prefer a slightly warm, humid condition with the temperature never dropping below 15 degrees (60°F.).

Arachnis flos aeris (Linnaeus) was originally named by the author *Epidendrum flos aeris*. Linnaeus, of course, lived in the eighteenth century and the history of *A. flos aeris* goes back to that period, when Kaempfer, a traveller and perhaps early botanist of sorts, named and actually illustrated the orchid as *Kunong ging* in the year 1712.

One of the genus has the peculiar habit of producing the lower flowers on the raceme which are of quite different colour to those appearing later. On *Arachnis lowii* the lower flowers, usually two or three of the many produced on well-grown plants, are light-coloured and spotted, while the others on the raceme are light yellow overlain with rich red-brown in broad patterns. The dimorphism is ill understood.

Arachnis annamensis (Rolfe) J. J. Smith was found by Sanders' collector Micholitz probably in much the same area in Indo-China where he collected *Cymbidium insigne* and *C. erythrostylum*. It was first flowered by Sir Frederick Moore at the Royal Botanic Gardens in Glasgow, where so many 'firsts' were grown and flowered.

Arachnis lowii (Lindley) was originally described in the *Gardeners' Chronicle* as *Vanda lowii* by Lindley in 1847. What is regarded by most as a variety, *Arachnis rohaniana*, appearing in Veitch's *Manual of Orchidaceous Plants* as the variety *rohdeniana*, probably was sent to England about the same time. Although there are significant differences in the two, they are sufficiently similar in habit to be mistaken one for the other.

Arachnis lowii was discovered by Hugh Low in Sarawak and sent to the nursery owned by his father in England about 1845. In cultivation and in nature the plant will grow to considerable height. They bore reputations as poor travellers and many were lost in transit by the slow methods of those years. However, several magnificent plants were recorded, among them some bearing many flower spikes, all distinguished by dimorphism, the habit prominent in the two almost separate species. *Arachnis lowii* variety *rohaniana* is never as robust as the other.

Arachnis clarkei (Reichenbach f.) J. J. Smith is subject to misleading information in some authorities, with the altitude of discovery quoted in one as 1000 to 2000 metres (3250 to 6500 feet) and in another as 2000 to 2700 metres (6500 to 8850 feet). The correct figure could preferably be taken from the *Manual of Orchidaceous Plants* as 1800 metres (6000 feet), and the environment and climatic inferences worked out from that as moderately warm and humid. It is too much really to infer environments incorrectly when growing species orchids such as arachnis, which at their best are rather chance subjects in less than moderately tropical climates.

In cultivation it is necessary to tie the plants to vertical stakes. These should be held in the pots with ballast rather than lighter material which may allow the pots to tip if knocked or accidentally brushed against. *A. clarkei* (J. J. Smith) flowers in ordinary good cultivation in late summer to autumn and it would seem that up to four flowers are usual. Others of the species have a similar number, allowing for the fact that the plants are not fast growing like other monopodials such as epidendrums.

Watering and fertilising follows that suitable for the culture of vandas, with the exposed roots the guide. In good conditions the growth period is during the warmer parts of the year, and there should be a dormant period when there is no activity at the tips of the roots.

Arpophyllum (La Llave and Lexarza) 1825

The *Botanical Register* described this genus as

> Some of the most graceful and beautiful of the Mexican orchidaceae recently introduced. The stems of *A. spicatum* are slender and a foot and a half high, with the sheathing as rough as shagreen leather; it is terminated by one long curved leaf, from the axil of which there curves in an opposite direction a dense spike of pink or pale purple flowers. It is a species of great rarity even in Mexico, where it has hitherto only been seen at Sultepec and near Arembaro, growing upon trees.

This is a quote from Paxton's *Dictionary of Botany* issued in 1868. Three species were listed and described — two from Mexico and one from Jamaica. La Llave and Lexarza were Mexican botanists.

Two species are recognised, *A. spicatum* (La Llave and Lexarza) and *A. alpinum* (Lindley), which he named in 1842. The latter grows at a higher elevation than *A. spicatum*, usually in cloud forest at about 3000 metres (9800 feet); *A. spicatum* grows at altitudes about half that height.

The collector who brought *A. spicatum* (Lindley) to Europe was Hartweg. But Roezl also sent plants to Europe and the orchids attracted much admiration. The genus was said to be in the pleurothalliae, but it is catalogued also as belonging to the laelieae.

Once thought to occur only in the Central American states, it is also found in Colombia, Venezuela and the island of Jamaica. There is some variation in the forms from different places, but in general it is easy to identify.

The genus grows always in bright light, principally on the tops of trees and rarely as a lithophyte in open situations. In cultivation bright light, rather more than is usually given to cattleyas, appears to suit the genus. This means it may be grown as a suspended plant close to a glasshouse roof where it is almost at risk of burning in bright sunlight. The narrow leaves can withstand this and become quite dark coloured.

Most cultivators will find that flowering period is spring to summer, but this may vary in some environments. The plants could be termed dry-growing epiphytes which will withstand flooding in the growing period and need adequate fertiliser to keep them moving. Usually the growths are erect, but if they tend to straggle they should be encouraged to grow erect so that the flower stems may grow vertically.

Growing mixes should be of coarse semi-terrestrial nature. The texture should be a little coarser than is usual for cymbidiums, so that moisture will penetrate quickly and the plants may be almost dried out as they cease flowering.

Arundina (Blume) 1825

A small genus of Asiatic orchids which extends down through Malaysia and associated islands. Now considered as a genus of only one member with the other species incorporated as varieties by some authorities. Originally collected and brought to cultivation in England from plants gathered in Burma and northern India, *Arundina bambusifolia* (Lindley) is now supposedly correctly known as *A. graminifolia* (D. Don) Hochr. However, the older name is more frequently attached to it and it is noted as such in the illustrations (page 108).

It has been known in cultivation for almost 150 years. Sir Joseph Hooker found it growing in the Khasia Hills, a region of northern Burma almost on the border with India and remarked of it: 'I know no more attractive picture of its kind than a patch of grassland. . . adorned with clumps of it in full flower.' The flowers open consecutively upward.

A. graminifolia (syn. *A. bambusifolia*) is more a tropical orchid and is well suited to gardens in those areas because it enjoys being undisturbed for long periods. Seen thus, with more than sixty stems, most of which are in flower, it is a most beautiful plant.

In cultivation and cool climates about the same temperature is needed as for phalaenopsis, but with none of the high humidity associated with that genus. Generally the stems of glasshouse-grown plants are much shorter than those of open-grown specimens, which are about 1 metre or more (3½ feet) tall, with the flower scapes above the heads of the growths. Flowering season is from late spring through summer in most instances, depending on the climate and the care with which the plants are grown. It may extend into autumn.

The species name indicates quite well the appearance of clumps, with slender bamboo-like growths, no pseudo-bulbs other than an enlargement of the base of the stem, and a close and prolific root system. It is considered to be related to sobralias, but unfortunately no hybrids appear to have been created. The flower resembles a cattleya, with pale pink to deeper coloured petals and sepals and a deep purple-red labellum.

When choosing potting material and pots, consider the preference of plants to remain undisturbed. While there is no reason why plants should not be grown in pots, this should be regarded as second choice and plants grown in the open such as a raised garden bed filled with some coarse material blended with half-decaying leaf-mould to which some peat-moss or polypodium fibre has been added. The plants resent dryness at any time, and when grown without adequate shade the leaves will discolour and probably burn and drop. In glasshouse cultivation about 50 per cent shade and clearance of 1 metre (3 feet) below the glass is advisable. Solid fertilisers are appropriate.

Ascocentrum (Schlechter) 1913

A small genus, most members of which were transferred from aerides and saccolabium. In some collections no doubt they are still known under the old generic title and also appear so in much literature.

As vandaceous orchids, ascocentrums are almost totally epiphytic. The choice for environments is limited and nearly all the genus need minimum climatic conditions above 12 to 15 degrees (54–60°F.), with humidity of 50 per cent or more and adequate light usual for most epiphytic orchids.

The colour of the flowers varies from the white varieties of some species to yellow through to purple reds. Try to provide the various species with light which suits the colours. For example, the bright *Ascocentrum curvifolium* (Lindley) or *A. miniatum* have better colour if flowered in almost clear light conditions unless it is bright direct sunlight. They flower in spring to early summer.

This genus is widespread over much of Indo-Asia, extending in the instance of *Ascocentrum miniatum* (Lindley) Schlechter, to the Philippine islands, found at various altitudes, but principally about 500 to 750 metres (1650 to 2500 feet) above sea level. This is about the altitude at which a great number of genera occur, and the environment is tropical to sub-tropical. This would be almost a constant factor for the genus throughout its distribution. For artificial culture, a warmed glasshouse would be advisable in naturally cool climates.

Cultivation of ascocentrums may be varied more than for other vandaceous orchids such as aerides. The root system, while open to the air naturally in wild plants, will accommodate to pots or wooden slat baskets. The choice of container will be influenced by the situation the plants occupy, whether hanging or benched. In either situation the container should be made more open by cutting slits about 2 centimetres (1 inch) wide in the sides of plastic pots. As these pots should be filled with coarse lumps of material, they may be larger than usual for epiphytic orchids. The roots will be found to inhabit the pots and send a proportion to act as aerial systems in a humid atmosphere.

Fertilisers should be used in cultivating any vandaceous orchids. Either liquid sprayed on the surface of the coarse potting material or solid fertiliser pellets applied in the growing season are equally effective. Note the amount of activity in the root systems and apply solid fertilisers, pellets and water only sparsely if the tips of the roots seal off.

Aspasia (Lindley) 1832

A small genus which has been in cultivation for more than 150 years. It was named by John Lindley in 1853 and the first species to flower in England was *Aspasia variegata* (Lindley), in the collection of a grower named Knight. It was a plant from Brazil, source of so many of the initial members of genera to be cultivated. The notification of its advent was in the *Botanical Register* of 1835–6. *Aspasia variegata* (Lindley) also occurs in Venezuela and Guyana and associated states and is recorded as growing in Trinidad, although rarely.

Aspasia lunata (Lindley), a Brazilian species and very beautiful member of the genus, was exhibited with many other species orchids by Sir Trevor Lawrence in the late years of the nineteenth century. The whole exhibit gained a certificate for the number of species and their rarity.

The genus is allied to the oncidium-odontoglossum complex, and it grows as an epiphyte in its habitats, which are widespread. Aspasias are seldom grown into specimen plants carrying many flowers because of the nature of the growth, which features a rising rhizome like some of the affiliates in oncidiums.

Aspasia epidendroides (Lindley) is probably the commonest species, occurring in widely separated habitats in Guatemala, Honduras, Costa Rica, Panama, El Salvador and down into Colombia. Most of the genus grow at fairly low altitudes of 300–400 metres up to 750 metres (1000 to 2500 feet) and higher. They need a moderate intermediate environment.

Aspasia pusilla (Schweinfurth) is a small Panamanian species which has been removed from cischweinfia, another genus. It is rarely noted.

The species frequently referred to as *Aspasia principissa* (Reichenbach f.) has come to be recognised as a variety of *Aspasia epidendroides* (Lindley). Unlike that orchid, however, *A. principissa* occurs in only some of the Central American states.

This last species referred to has an importance in its use with other genera, principally the odontoglossum complex, in cross-pollination. This can produce flowers which are similar to odontoglossums but will tolerate warmer conditions.

In cultivation as epiphytes, use has been made of pots and slat baskets, with the latter offering better aeration of the open root systems. Potting material can be any type, from fibres to coarse bark, which offers dryness when necessary for dormant plants.

Baptistonia (Barbosa-Rodrigues) 1877

A genus of a single species little known and cultivated in the late years of the 20th century. The species *Baptistonia echinata* (Barbosa-Rodrigues) was originally named *Oncidium brunleesianum* (Reichenbach f.) in 1883. If ever an orchid has hidden its light under a bushel it is this species.

Native to Brazil, it is known to have grown in only two localities, Rio de Janeiro and Sao Paulo, although some authorities give other derivations for it. As an oncidium it was recorded in Veitch's *Manual of Orchidaceous Plants* and its affiliation with the oncidium genus was brought into doubt at various times in the years of its first introduction. It also had at one time indications of inclusion in the genus epidendrum, but that was a common designation for many orchids later removed to alternative genera or new ones.

Its history as related in Veitch's book is interesting:

> In 1883 Mr W. B. Lemon, of Moat Lodge, Beckenham [England], obtained through an engineer, of Rio de Janeiro, a small parcel of mixed orchids, not necessarily collected at Rio, among which was an Oncidium resembling *Oncidium sarcodes*, but which upon flowering proved to be a very different species. In 1885 it was awarded a First Class Certificate by the Royal Horticultural Society, and in February of the following year it again flowered in Mr Lemon's collection, producing a densely branched panicle of upwards of 150 flowers . . .

Alternative sources of the plant were brought forward, but with the state of information at the time in the late years of the 19th century some doubt was cast on the truth of the tales.

Plants of the species have the attributes of oncidiums: the leaves and pseudobulbs conform well to the morphology of that genus. Treatment of the plant follows that given oncidiums and, provided dormancy and activity are noted and the moisture and light adjusted accordingly, there should be little problem getting it to flower in the period following maturity of the bulbs. It is doubtful, however, whether the number of flowers could be persuaded from a plant to equal those of Mr Lemon's plant.

Shade is always a consideration for oncidiums and similar genera. While the intensity of light should not be such that it could produce a suggestion of burning, particularly on immature foliage, it should be enough for the leaves to produce a light green colouring rather than the lush soft green that too heavy a shade produces.

Barkeria (Knowles and Westcott) 1838

A genus which, although discovered in the early days of orchid growing, was never freely cultivated. It was removed from its first name and incorporated as an epidendrum. However, it was reinstated by Thien and Dressler in 1970 and described in detail by Halbinger. Leon Wiard, in his book *An Introduction to the Orchids of Mexico*, lists the four species occurring in that country, and others are recorded in various publications which detail the orchids growing in Central American countries.

Barkerias were one of several genera known under the title epidendrum, another notable instance being early cattleyas. The Mexican species are *B. chinensis* (Lindley), *B. dorothea* (Halbinger), *B. barkeriola* (Reichenbach f.) and *B. elegans* (Knowles and Westcott). Other states of the region also have representatives, including *Barkeria lindleyana* (Bateman ex Lindley), which is listed as common to all the states of the Central American region by some authorities. The altitude at which the plants grow varies from about 800 to 2500 metres (2600 to 8200 feet).

The genus is epiphytic and occasionally lithophytic. *Barkeria elegans* (Knowles and Westcott) was the first species introduced into Britain in 1837; it was subsequently lost for a number of years until reintroduced by Linden in 1853. No doubt the reason for its loss was that the plants' needs were not understood, as was the case with many other orchids of the period. *B. elegans* comes from the mountains on the Pacific coastal regions of Mexico, and has always been a popular cultivated plant.

It is probable that taxonomy will deal with the genus again because of its similarity to some of the epidendrums. An instance is *Epidendrum stenopetalum* (Hooker). He named that species in 1835. It is here described under epidendrums (page 68) and illustrated on page 138.

The morphology of the genus varies from short, laelia-like pseudo-bulbs to rather spindly growths to about 35 or 40 centimetres (16 inches or more), the flowers appearing on the apex of the pseudo-bulbs in late autumn. Leafage is impermanent and once the flowers are finished the plants move into dormancy and should not be watered as freely as when in growth or flowering. Temperatures should not fall below 10 to 12 degrees (50–55°F.), and range upward into the usual summer temperatures in sub-tropical zones. The genus as a whole is unlikely to acclimatise to tropical climates.

Bifrenaria (Lindley) 1833

Lindley, when naming this genus, was considering one of the earliest South American orchids brought to cultivation. It is a genus of about twelve species, most of which are from Brazil, the centre of origin of so many of our cultivated orchids.

Probably the best-known species is *Bifrenaria harrisoniae* (Lindley), which Hooker named *Dendrobium harrisoniae* in 1826. It was cultivated as an exotic plant much earlier than that in the collection of Mrs Harrison, who originally came into possession of it through a relative in Rio de Janeiro. Mrs Harrison was recognised at the time as having a notable collection of orchids and she grew them well.

Some of the genus have peculiar pseudo-bulbs, almost square in section; this is a characteristic of other orchids, including dendrobiums. The pseudo-bulbs of most bifrenarias are rather squat, tapering slightly, pear-shaped, toward the apex and usually carry a single leaf. It is common also to find that the pseudo-bulbs become ribbed as they age and do not fill out again as those of other orchids do.

Bifrenaria aurantiaca (Lindley) was removed from the genus by Rudolf Schlechter and named *Lindleyella aurantiaca*. It was again renamed by Hoehne in 1944 as *Rudolfiella aurantiaca* (Lindley) Hoehne, and is beautifully represented in colour in his *Flora Brasilica*. It carries a spike or raceme of golden-yellow flowers freely spotted with red. *Rudolfiella aurantiaca* is also known in Guyana and rarely in Trinidad. It was noted particularly in Guyana by the Schomburgk brothers.

Bifrenarias produce albino varieties, although not so prominently as other Brazilian genera. Albinos can occur in *B. harrisoniae* (Lindley), a very hardy species and the commonest found in collections. It is usually recommended for bright conditions, a warm environment with temperatures which do not drop too low in cooler periods and a positive dormant period of some months when it neither grows nor flowers. However, good and free flowering has often been noted on plants which have leaf colour indicating that they have grown in shaded conditions. If the plants are suspended where they will not be too frequently watered and the leaves become at times over-exposed to light and turn slightly yellow, more frequent flowering will result. Wasting of the pseudo-bulbs should be controlled and plants not allowed to deteriorate too far. Flowers usually appear on the newest pseudo-bulbs.

Other bifrenaria species which give an equal and occasionally better array of flowers include:

B. inodora (Lindley), Brazilian, principally green and somewhat similar in shape to *B. harrisoniae* (Lindley). Cultivation is similar, with a definite dormant period through some of the annual cycle.

B. maguirei (Schweinfurth), another Brazilian species, named by Schweinfurth in 1959, is unlike the two we have discussed. It has long slender leaves on a petiole of half the total length; the spike or raceme is taller and carries two or more flowers. It is cream to yellow in colour. This bifrenaria also grows in Venezuela, where it was found at about 1500 metres (5000 feet) thriving in boggy situations, which is unlike the conditions which most of the genus prefer.

B. tyrianthina or *thyriananthina* (Loddiges). The spelling of the name varies, even in the one publication. This Brazilian bifrenaria is pinkish with a blue tint. The colour is difficult to describe, but the general habit of the plant is much the same as *B. inodora* and the flowers about the same size, some 4 centimetres (1½ inches) across.

B. wittigii (Reichenbach f.) is cream with a dark labellum and about the same size as *B. tyrianthina*. It is solely Brazilian, the flowering habit as in *B. harrisoniae* or *B. tyrianthina*.

B. longicornis (Lindley) is known as native to Guyana and Peru, where with another species, *B. saxicola* (Schlechter) Schweinfurth, it makes up the only two bifrenarias native to that state.

Bifrenaria saxicola is also known as *Lindleyella saxicola* (Schlechter) according to some authorities. With the general taxonomy of South American orchids at times confused it is best to keep to the names under which the plants are procured.

In general the genus can be grown in well-drained cymbidium mixes, but consistent with the description of the habitat of *B. maguirei* (Schweinfurth) no positive system can be lined out for the total genus. They should be grown in systems which allow a certain amount of flexibility. For those grown in pots, choose a medium which can be dried out or remoistened at will, consistent with the condition of the plants and their known habitats.

All the genus may be cultivated in intermediate conditions where the temperature does not fall below about 12 degrees (54°F.) in the cooler period of the year, which is when plants are usually dormant.

Fertilisers should preferably be inorganic liquids which may be boosted with quick-decay solids, applied only in growing and flowering periods.

Brassavola (Robert Brown) 1813

A widespread genus, native to Central and South America and some Caribbean islands. It extends as far south as Bolivia and Paraguay. Probably the commonest of the genus is *Brassavola nodosa* (Linnaeus) Lindley. Linnaeus, the original author of the species, named it *Epidendrum nodosum* in 1753.

There are about fifteen members in the genus, some of which were dealt with in *Growing Orchids, Book Two*. Strangely, the genus was not mentioned in Veitch's *Manual of Orchidaceous Plants*, although some species have been grown throughout the history of orchid cultivation. Some species are delightfully fragrant, particularly at night.

Most have terete foliage (see page 6), some of which is rather deeply channelled lengthwise down to the petiole or short stem connecting the leaf to the creeping rhizome, from the joint of which stems emerge, carrying two or three flowers. The bases of the leaves are covered with a papery sheathing system.

The genus generally should be considered as semi-tropical to dryish tropical and preferably grown open rooted. The plants are not suited to container culture except in very open, coarse mixtures based on bark. Some grow in almost desert-like environments in company with xerophytic plants such as cacti, principally as epiphytes on stunted trees and infrequently as lithophytes. They should be grown in an open, airy position, perhaps above other plants in enclosures. They do not like the humid conditions associated with high temperatures which are provided for phalaenopsis or some vandaceous orchids.

One of the earliest species introduced to England was *B. cucullata* (Linnaeus), brought from the West Indies by Vice-Admiral Bligh, an officer in the British Navy whose name is well known in the history of the Pacific. It was illustrated in *Growing Orchids, Book Two*.

Eight brassavolas are listed in *Orchidaceae Brasilienses*, by Pabst and Dungs, including *B. ceboletta* (Reichenbach f.), which first appeared in Schiller's collection at Hamburg, Germany. Like other species in the genus, it had a confused history, although widespread in Brazil and other South American states. The genus as a whole grows at low to medium altitudes.

One of the best dissections of the genus, perhaps incomplete, is the summary by R. A. Rolfe in the *Orchid Review* for March 1902. At that time, of course, *Rhyncolaelia digbyana* (Lindley) and *R. glauca* (Lindley) were still included in the genus and described by that author as 'a most curious orchidaceous plant obtained near Vera Cruz [Mexico] for the Horticultural Society by Mr Hartweg.' Others in Rolfe's period have been reduced to synonymy.

Older records show that the genus was never as popular as many of its associates, such as laelias and cattleyas, from the South and Central American states. This can be understood when the cultural methods and establishments of the time are appreciated. But in modern orchid growing in good environments and with the correct associates the genus as a whole should not present too many difficulties.

Any of the species are most rewarding plants to grow because over the period when there is no growth and no flowers they may be hung out of reach of water and merely looked over periodically to see that pests such as scale do not lodge in the bases of leaves. Overall there is much similarity in the flowers because most have a white labellum. It is their outstanding feature.

Plants may be propagated by division, and during removal the root system is such that some damage will necessarily take place. It is as well to understand the time when growth should be expected so that the plants can be brought back to vigour by careful watering. Too much water at the wrong time soon kills off the root system and then a fresh start must be made.

During growth and flowering, liquid fertilisers should be sprayed over the root system and the growing base of the potting material. Root systems seldom go deep and are usually rather straggly. The foliage, being terete, gives all the indication for protection needed. Shade should never be dense.

Plants may be grown in pots or in slat baskets. The former should have four openings in the sides about 1–2 centimetres (up to ¾ inch) wide and as long as the strength of the pot will stand. It should not be necessary to line the interior for such coarse-rooted species as brassavolas but, if it is felt that too much material will be lost, a liner of shade cloth can be used.

Slat baskets are very easily made, but do not use treated timber or resinous pine. The basket may also be lined with shade cloth. Some growers make bags of shade cloth and use them for many kinds of orchids. Any epiphyte which needs a part or fully open system has much scope in this type of container. Genera which need periods of dehydration are well suited, but some attention is needed to see they are frequently moistened.

Brassia (Robert Brown) 1813

A Central and South American genus closely allied to miltonias as distinct from miltoniopsis, the broader labellum of which distinguishes this Colombian orchid from the more durable Brazilian genus. The flowers of the brassias are in most instances rather spidery on slender semi-erect spikes and usually well spaced and displayed. Not all are large like those illustrated in this book.

Probably the commonest in cultivation is the principally Central American species *Brassia verrucosa* (Bateman), although others of the thirty-odd members of the genus are also popular. There has been confusion over names, partly because of doubling of registers in different countries. Charles Schweinfurth's Peruvian catalogue of the genus has the footnote in an otherwise careful and well-authenticated list that material was not available and reliance placed on drawings or photographs.

Perhaps it is well to note here the confusion which occurred in the registrations of at least one authority — Professor Reichenbach, usually denoted in relation to various species as Reichenbach f. In his lifetime he described a very large number of orchids, specimens of which were sent to him from all over the world, particularly Britain. This material he used as the basis of a large herbarium, which was to be made available to the horticultural world on his death. Instead he locked it away for twenty-five years following his death in 1889. His will noted that: 'My herbarium and my botanical library, my instruments, collection of seeds, &c., accrue to the Imperial Hof Museum in Vienna. . . [and] shall not be exhibited before twenty-five years from the date of my death have elapsed. . . .'

Those wasted years meant that much material was reprocessed and re-described and there was no recourse to the authority which had so jealously guarded it. Some of that confusion certainly attached to brassias.

In May 1914 Reichenbach's embargo was lifted and his life's work was released for study. It was contained in fifty-nine cases, of which eight concerned orchids. In all, his notes and manuscripts were said to comprise 420,000 pages. It was apparent from a summary of his notes published in the *Orchid Review*, that his task got beyond him. The year 1914, of course, saw the outbreak of war which again had the effect of sealing up his work and in the four years that followed little work was done. As we move toward the close of the twentieth century there is still much to be done.

We return to the brassias. In growing these orchids either of two methods are successful: the plants may be potted in coarse material or attached to vertical pieces of treefern or wood.

The genus generally appears to prefer open-rooted cultivation and produces masses of aerial roots which do not enter pots or material but attach to suitable vertical supports. Those grown in pots tend to overgrow the containers and prefer aerial root systems. Where plants overhang the edges of containers they produce flowers prolifically in their season, which is usually summer, occasionally early autumn. Intermediate to warm conditions suit the genus, which grows at all elevations up to some 2000 metres (6500 feet). Two brassias have been transferred to the genus ada.

Brassias were detailed in *Growing Orchids, Book Two*, but *B. caudata* (Lindley), a prominent Brazilian species, and several other lesser-known species such as *B. angusta* (Lindley) and *B. lanceana* (Lindley), which extends its habitat to Venezuela and Guyana, were not included. There are thirteen species in the Brazilian catalogue, some of which are infrequently known and grown.

Although the genus may be cultivated on benches, it is more successfully grown and managed as suspended plants. This may require the strengthening of roof members, or the plants may be hung on mesh of 5 to 10 centimetres (2 to 4 inches) which is suspended vertically in glasshouses or enclosures. This type of cultivation is economical on space and has the advantage of providing a shady side and a bright side if the sections of mesh are correctly placed.

There is no great variety in colour as in some other genera, the greatest amount being the heavy blotching in the red-brown areas in *Brassia gireoudiana* (Reichenbach f. and Warscewicz), native to Costa Rica.

Brassia caudata (Linnaeus) Lindley is a representative which grows in the North American continent in southern Florida. It grows in various habitats and is frequently known by what could be regarded as a synonym, *Brassia longissima*.

Fertilisers most commonly used for brassias are liquid and should be applied only when root tips are growing. The growing and flowering seasons are closely related and in some species the growing and maturing is done in the cooler months of the year in artificial culture.

Temperatures should be kept in the intermediate to warm range. Brassias may grow in cooler conditions but possibly will not flower. Minimums above 10 degrees (50°F.) are recommended.

Broughtonia (R. Brown) 1813

A genus of only one member, although some literature still quotes *Laeliopsis domingensis* (Lindley) as a broughtonia. This species also has been known as *Broughtonia lilacina* (Henfr.), under which name I cultivated the plant in my early inexperienced years. *Broughtonia sanguinea* is localised in the Caribbean, particularly on the islands of Cuba and Jamaica, but its presence in other countries of the area cannot be discounted. The plants are hard-looking and will tolerate most adverse conditions of bright light and warmth. Good conditions and lighter cultural techniques rather modify the hard appearance of the plants and they can become softer looking and produce better flowers.

Broughtonia sanguinea (Robert Brown) was one of the first exotic epiphytes cultivated in England, when it was grown in Kew Gardens in 1793. It was described as a bright crimson-purple, and one of the apparent tints noted in the flowers is still that indefinable purplish background colour. (It is illustrated in *Growing Orchids — The Specialist Orchid Grower.*)

While some credit is given for it as lithophytic, sometimes referred to as rupicolous, it is principally as an epiphyte that the species is recorded. Altitude in the island habitats appeared unimportant as it was found variously from sea level to 2000 feet. It was often found in company with *Brassavola nodosa* (Linnaeus) Lindley.

Cultivation of broughtonia is simpler in warmer areas and preferably with low humidity for much of its growing and flowering periods. After they have flowered and are apparently dormant the plants can easily be damaged by frequently applied moisture. However shrivelled its appearance, a plant soon picks up moisture once root or vegetative growth appears. Do not rush to water it even at this stage, as the fattening out and flowering of growth in spring to summer follows normal culture.

Growing techniques vary from grower to grower. While some may prefer open-root culture on pieces of fibre or suitable hard wood or cork slab, when this proves unsuitable to the root system it is better to replant the species on a changed host as soon as new roots appear at the base of new growth or on the older sections of the system. This usually follows a change in season and increased humidity in the habitats.

In collections of genera, some species will not fit into a common pattern necessary for most artificial environments. Broughtonia is apt to be non-conformist and it may be difficult to keep plants in continuous health. It may be necessary to discard the genus if it is ill suited.

Bulbophyllum (Thouars) 1822

A genus usually considered the most numerous of the orchid world, containing according to some authorities more than two thousand species. It is also one of the most widespread, with members in Indo-Asia, the Philippines and down to the Indian Ocean islands, Africa and its associated land masses in the southern sector, also its western states, Australia and across the Pacific Ocean to Central and South America.

In its widespread habitats the morphology of the group remains fairly constant, although there are great differences in the shape and size of the inflorescences. The largest plant form is usually considered to be the Borneo species *B. beccarii* (Reichenbach f.), although the individual flowers are quite small. A footnote in Alex Hawkes's *Encyclopaedia of Cultivated Orchids* says that the stench of the inflorescence could be 'likened to a herd of very dead elephants'.

It is difficult to say which area has the densest population of bulbophyllums, but Niugini and contiguous islands and Irian Jaya could contain more than five hundred species.

Although not a popular genus in cultivation, some of its members are very beautiful and most intricate in their floral formations. Its relative importance through the history of cultivation is also minor and a review of early literature on orchids does not reveal any of the emphasis given other genera, particularly in the splendour of the collections of late years of the 19th century.

The four-volume set of Frederick Sander's *Reichenbachia* does not include one bulbophyllum species. The genus does not appear in other books of lithographs portraying the flowers of the period. A few were included in Curtis's *Botanical Magazine*, some of which was reprinted in 1986.

There are many spectacular flowers on plants of the genus, one of which is *B. ericssonii* (Kraezlin). This Niugini species was freely cultivated in the late nineteenth century and some magnificent specimens are recorded in the lists of the Royal Horticultural Society in England. Another most unusual flower is carried on *B. medusae* (Lindley) Reichenbach f., from Malaysia and Borneo.

The smallest orchid in the whole world catalogue must still be considered as *Bulbophyllum globuliforme* (Nicholls), but there are other species almost equally minute. The pseudo-bulbs of the plant I have grown for fourteen years measure about 1 millimetre (1/25 inch) in diameter. But let us consider a contender.

In the *Orchid Review* of June 1915, the editor, Robert A. Rolfe, included a short article on minute bulbophyllums. He commenced with a review of the history of *Bulbophyllum minutissimum* (F. Mueller), perhaps not knowing that it had an even smaller close relative in *B. globuliforme* (Nicholls).

The second part of the article dealt with what in 1880 was called the smallest orchid in the world, found in Borneo by Beccari, which was smaller than *B.*

minutissimum. These two were morphologically similar and the Borneo species was named *Bulbophyllum odoardi* (Reichenbach f. and Pfitzer). From the description it is apparent that it was perhaps larger than *B. globuliforme* (Nicholls) but perhaps smaller than *B. minutissimum* (F. Mueller). It is debatable. But those who have seen *B. globuliforme* in flower will appreciate that for anything to be much smaller it would be necessary to look at it with respect and a magnifying glass. *B. globuliforme* is illustrated on page 115 together with a red-painted head of a pin to give an idea of the size of the pseudo-bulbs and the flowers.

In a general sense the growth habits of bulbophyllums are similar from species to species. The appearance is of a creeping rhizome with, along its length, pseudo-bulbs which are principally single-leafed. These pseudo-bulbs begin as extensions of the rhizome from the bases of older pseudo-bulbs and may occur at any point along the length of the plant provided the pseudo-bulbs retain foliage.

The distance between the pseudo-bulbs varies and the flower stems may emerge from any of the brownish scales or bracts which cover the rhizome. In some of the species, notably the Australian genera which creep over surfaces, these bracts or scales may flower for indefinite periods but they are always seasonal from early spring in cultivation.

Considering the habit of the plants and the nature of the root systems, flat dishes or pans are most suitable and the growing material should be coarse. The root system emerges from the lengths of rhizome and generally not from the base of the pseudo-bulbs. It is short and reacts to overwatering. The plants are epiphytic and lithophytic and the root systems must be treated as open, requiring no great depth of potting material.

Fertilisers should be used sparingly and the habits of the species understood individually, plant to plant and species to species. Note should be taken of the growing habits and the time or season at which the plants initiate growth. It will be found principally to occur in late spring to summer.

The genus usually prefers warm to intermediate climates, and very few grow in cool to cold conditions.

Calanthe (Robert Brown) 1821

A broad-leafed terrestrial genus with widespread habitats in Central America, the Caribbean islands and Pacific countries, with the best-known members from Indo-Asian countries and one member in Australia. The genus contains the member species which were used to produce the first recorded orchid hybrid, *Calanthe Dominyi*.

As indicated by their foliage, calanthes are shade-loving plants and they have some aptitude for attracting insect pests known also in other genera. These include two of the banes of artificial orchid culture, red spider and mealy bug. Both these pests are widely described in orchid literature and it must regrettably be stated that they can be dealt with best by use of fungicides and sprays that are both poisonous and undesirable in any context.

Most of the species growing naturally in the Indo-Asian region have been cultivated more frequently than those from other areas and it is among these that cultivators find the material for their collections. *Calanthe vestita* (Lindley) has many original habitats stretching from Burma across Indo-Asia to the Celebes Islands and the Philippines in the western Pacific Ocean. Considering these origins it is not surprising that the genus generally is cultivated more in tropical and warm climates than in cooler regions where artificial heating is necessary. The use of heat tends to shorten the inflorescence of most species.

Calanthe vestita (Lindley) and *Calanthe rosea* (Lindley) Bentham were found growing together by Boxall in Burma and with them an obvious intermediate form which indicated that the two had cross-pollinated. This cross-pollination was also made by Dominy, the seed grown and finally flowered and named *Calanthe* Veitchii. This, of course, raises a question which may well be asked about many species orchids cultivated by growers and including some Australian orchids. Are the natural hybrids to be regarded as species in their own right? Of course, to all intents and purposes they are simply a movement through stages which can be regarded as steps in evolution.

Calanthe veratrifolia (Robert Brown) originated in a very wide arc which stretches down into the Pacific islands and Australia. In the latter it grows in semi-tropical to sub-tropical and even temperate climates, and it can be grown in moderate conditions in artificial cultivation.

Some of the species shed their foliage almost as a preliminary to flowering in autumn and winter. In artificial conditions they must be cultivated in climates which promote fast growth to be completed no later than autumn. To produce these results it is best to repot the leafless pseudo-bulbs as soon as new shoots begin at the base.

Potting material should be rich in humus and free draining so that it does not become waterlogged. The plants should not be allowed to dry out until the leaves have been shed and the flower spikes are well grown.

The shape of the pseudo-bulbs in *Calanthe vestita* is curious. In the mature state they have a marked two-part appearance, as though they have had a restricting ring tied about the top one-third of the pseudo-bulbs.

Not all the genus is deciduous and *Calanthe veratrifolia* (Robert Brown) is among these. The foliage develops brown spots if the plant experiences a drop in temperature after watering. The spotting is not confined to this species and is cultural rather than a natural phenomenon.

Pests should be kept under control; the plants must be constantly watched and prophylactic or killing agents applied as necessary. The usual control for red spider is moisture, but as this tends to mark the leaves some other means such as pest strips or retardant sprays under the leaves must be tried. Pest strips do control many pests in enclosures and are worth using for any genus. Their recommended life may be reduced by as much as half in humid environments.

Calanthes will take frequent applications of fertiliser. As the root system is mostly renewed each year, the only precaution to take is to not over-water or over-fertilise them. Inorganic or organic liquids are better than slow-release tablets or pellets. The only pelleted fertiliser recommended (unless good experience dictates otherwise) is poultry manure pellets; these are sterilised and any residual bacteriacide diminished.

As indicated by all orchids with soft-looking green foliage, shade should be denser than for such orchids as cattleyas; careful positioning of plants in enclosures is the best way to work out this factor.

Annual repotting and the need for minimum temperatures of 12 to 15 degrees (54–60°F.) make the genus more suitable for warm climate cultivation than cooler areas.

Catasetum (L. C. Richard ex Kunth) 1822
Cycnoches (Lindley) 1832
Mormodes (Lindley) 1836
Chysis (Lindley) 1837

The first two genera are among the curious botanical specimens seldom occurring in the 1200 or more genera which make up the orchid flora. They bear either or both gender of flowers on the same plant in the same or successive years, but seldom on the one inflorescence. Mormodes and chysis are grouped with the others because their cultural needs and habitats are similar.

Confined to the Central and South American tropical regions, catasetum bear little resemblance to most of our cultivated orchids. None of the four genera attract cultivators in numbers usual to communities which grow principally cattleya-based hybrids or species collections, and there are good reasons for this. It soon becomes apparent that the four genera are almost a race which must be grown separately, with a lot of attention to detail. Much of this detail does not fit into the cultural patterns evolved for growing our main genera.

In cultivating insular groups such as phalaenopsis or vandas, to list only two genera, it is best to form collections of, say, fifty or more plants and to be particular in choosing genera to accompany them in select environments. This brings into question the value of such genera as catasetums, cycnoches, mormodes or chysis as import and quarantine items. Supply of such genera is limited and it is almost impossible to have plants guaranteed correct as to name. Considering the period over which these genera have been known and grown, as indicated by the dates of naming of each genus, the catasetums in particular are still improperly named and authenticated. It is not proposed to repeat in this book all the detail of their history, which is given in *Growing Orchids, Book Three*. However, some cultural ideas are necessary.

There are more than a hundred catasetum species, ten or more cycnoches, and about twenty mormodes. Their flowering habits differ, with catasetums flowering from basal or higher sections, cycnoches from the upper parts of the plants and mormodes from lower segments. The plants of all are morphologically similar and very close in appearance also to chysis.

Chysis and mormodes each bear perfect flowers and never have separate male and female flowers. Catasetums and cycnoches, as we have seen, bear either or both gender, in each instance with separate and distinct character and appearance. This is explained and illustrated in *Growing Orchids, Book Three*.

Each genus originates in a particular environment: nearly all the Brazilian catasetums, for instance, grow in bright locations with ample air flow about the plants and constant humidity. Some species grow on palms in the butts of old and fallen fronds, finding moisture which may remain in the butts for some time. Other species grow in the higher branches of light forest and in bright light. All are epiphytic and seldom lithophytic.

It is possible to grow individual plants in conjunction with other genera, but they would need some careful consideration. Temperature in both growing and resting stages would need to be constant and relatively high, never falling below about 15 degrees (60°F.). It is possible they could succeed when grown as upper level plants with phalaenopsis, but some concessions would need to be worked out between these genera so that, while neither was absolutely suited, they would both grow and flower. Other possibilities include paphiopedilums, with similar temperature restrictions and good light; or hard-cane dendrobiums, where still more adjustments would need to be worked out to get a little of everything for both types of orchid. Controlling humidity is important: it should not be too high.

In dormancy, which is distinct for all four genera as they cast their foliage, the plants should remain basically dry, with very little water applied infrequently to the potting material. Most of the root system will not return to activity and some on the newer parts of the plants must take over until new roots develop on new growth. None of the four genera flower from older pseudo-bulbs. The scope of new growth to produce flowers varies. On catasetums, cycnoches and mormodes it appears as inflorescence on almost or totally mature new growth. On chysis the flowers appear on very immature new shoots from the last-made growth, occasionally fairly high on this growth. After flowering, these growths mature and in this process occasionally some leaves will be cast from lower segments.

The four genera should not be grown as benched plants: they will grow and flower much better if grown in suspended containers, either pots with slots in the sides or slat baskets. The material used in these should be coarse and not capable of holding much moisture at all. The root systems vary but generally could be considered medium to fairly coarse.

Fertilisers are necessary for all artificially cultivated plants. In correct balance such plants as any of these genera should show increasing size of pseudo-bulbs and in the best of circumstances a double-break to increase the annual growth from one to two or more new pseudo-bulbs.

The ideal aspect for any of the four genera is facing the brightest light available and suspended fairly high in glasshouses or enclosures. As with any orchids, their culture should be planned. Do not add them to your collection simply because of their attractive flowers or as part of the general challenge to grow all genera or species.

Few people would consider these four genera as a total basis for orchid growing. Plants are difficult to obtain, and their requirements are strict, which makes it hard to add other genera. This applies most particularly to the species grower as such, because there is much attraction in the variety available in these projects, particularly in cooler climates.

There are combinations of plants not mentioned here, but a primary consideration should be the type of environment in which they originate and possibly the availability of genera.

Each of the genera is capable of subsisting for some time in poor condition, but to regard inevitable discard or death for weak plants is too much to contemplate with anything like complacency in view of the real decline of numbers still growing in their habitats.

This rarity brings into focus another factor. Each grower who has catasetums or cycnoches should have at least two plants of each species. These two should be grown in different modes and encouraged to flower. It is possible to encourage the two genders to appear in this way and if possible the plants should be cross-pollinated, each way if by a remote chance they both produce the requisite flowers. Cross-pollination is much better than self-pollination. It is better to make sure of one seed capsule rather than trying to over-produce capsules and so strain the plants.

If the plants set capsules they should be watched carefully and kept dry. The sign that they are about to split and shed the fine seed is slight discolouration. Arrangements should be made to have the plants raised by someone who knows what they are about. It may be preferable to remove the capsule before maturity and open it in sterile conditions in order to collect the seed.

Species orchids of these four genera and, in fact, any orchids from the Central and South American regions should preferably be handled and grown in this way and an effort made to secure their future in cultivation.

Cattleya (Lindley) 1824

It is difficult to find words to describe the magnificence of these South and Central American orchids. It is a complex genus about which there are disparate views. The real history of the *Cattleya labiata* complex was lost in the secrecy and deception surrounding discovery and collection of the various species. Discrepancies of hundreds of miles and much other misinformation was purposely disseminated about the habitats, each group or collector seeking to preserve their monopoly for as long as possible.

The genus as a whole can be roughly divided into two: the unifoliate or single-leafed species and the bifoliate cattleyas. Of the two groups, the bifoliates have smaller flowers and the unifoliates the most colourful.

The original unifoliate cattleya, *C. labiata* (Lindley) is still credited by many authorities as the true single species, and most of the other similar species — excepting the natural hybrids — are considered as varieties of it. There is some validity for both views. The facts were carefully considered by A. R. Rolfe, the first editor of the *Orchid Review* and one of the brightest analytical minds of all time in orchidology. He found it difficult enough trying to solve the origin of the first plants, but his opinion is probably the best we have.

The extent of variation in *Cattleya labiata* (Lindley) is exemplified by the publication in the *Orchid Digest* (separately reprinted) of a summary and many coloured illustrations. It is not too much to say that some of these variations may be wrongly identified.

Some information has been given on the morphological differences certain plants exhibit, but these again are unreliable because of the way changes occur in cultivated plants. The flowering season of the various species as given in some publications also are not a good guide, but in reality all can be verified. In cultivation there is much variation.

Frederick Sander, writing in *Reichenbachia*, had this to say of *Cattleya mendelii* (Backhouse) in 1888:

> *C. mendelii* is a comparatively new orchid, since it was only a few years ago that the first living plants of it were received by Messrs Backhouse, of York. Its home is in the Andes, north of Ocana, where it grows at an elevation of about 3500 feet [1000 metres]. Immense quantities have been imported to Europe, so that now it is sparingly found in a wild state. It is regarded by many as the loveliest of all the varieties of *C. labiata* or even of all Cattleyas.

Two parts of that brief quote should be noted. First that its presence in the wild had been reduced to rarity in so short a time, and second that it was regarded as a variety of *C. labiata* (Lindley). It was illustrated in *Growing Orchids, Book Two*.

It is not proposed to give lengthy descriptions of the various species, but to list a few here to familiarise new growers of species orchids with their names and flowering seasons. In my experience with these species, some should be regarded as not true to their name tags, but in general they are mostly propagations of plants which have been in possession of growers for some time.

The total number of cattleya species is about sixty, and the majority are bifoliate types. These bifoliate cattleyas extend more fully into Central America than the labiata section. They are in general taller and more slender, and some carry racemes of flower with greater numbers than the unifoliate group.

Among these bifoliates is possibly the first cattleya known to the Western world. It was *Cattleya loddigesii* (Lindley), first known as an epidendrum very early in the nineteenth century. *Cattleya loddigesii* is a cool-growing species from Brazil and Paraguay and is among the very few of the genus which will tolerate low temperatures. Its flowering season is usually autumn. In this way the growth period is over the warmer part of the year and the coldest part is spent in dormancy.

Other bifoliates are principally warm-growing and should not be artificially cultivated in temperatures below 12 degrees (54°F.). The principal distinction in this group, in comparison with the unifoliate cattleyas, is that the flower segments are mostly uniform in width and the labellum inclined to be spade-shaped. *Cattleya walkeriana* (Gardner), which may at times be unifoliate as well as bifoliate, exhibits this labellum shape to perfection. It is not among the illustrations, but is a Brazilian species, warm-growing, illustrated in *Growing Orchids, Book Two*. Some list it as cool-growing.

The nomenclature of many cattleyas is in dispute between various authorities and this disagreement is not uncommon in the orchid-growing community generally. Changes in spelling are frequent.

Specialists who choose to grow cattleyas will find that they are good mixers in a varied collection, growing as epiphytes in minimum temperatures of 12 to 15 degrees (54–60°F.). They are orchids which may be grown with medium protection in all but tropical climates.

Unifoliate and bifoliate cattleyas may be cultivated as epiphytes associated with many other genera. The choice of accompanying genera should be based on compatibility in environment: the plants should be capable of growth with moderate light and a good air flow. It is advisable to introduce external air rather than rely on internal circulation such as satisfies phalaenopsis and some of the vandaceous orchids.

If the plants are to grow on benches, the bench-top should be of an open form, rather than closed benching with a solid top. In this way an air flow is created upward through the plants. Easy alternatives are to build frames and hang the plants from them, or make a system of A-frames where the plants are held against open mesh sheet. Either system gives the plants good air circulation.

Flowering season of the species varies from autumn to spring, with very few flowering in summer. This last season is in general the growing period for all species and only a few plants flower in it. With varying flowering seasons the habit of each species should be understood and the plants segregated as they mature and flower. Some plants will mature and then hold the flowering sheaths for a period before the flowers appear; fortunately they are few.

The root system of all cattleyas must be understood. Its activity is seasonal and is indicated by growing root tips which are mostly green but occasionally may be darker coloured. When the root tips become white and seal off, it is time to reduce the amount of moisture applied to the plants without sacrificing humidity, which should be held at 40 to 60 per cent.

The usual potting material is based on bark, with very little addition of other elements. Occasionally charcoal is added, but unless there is good reason to include it, a mix entirely of bark will be found adequate. Containers should have the propensity to aerate following watering and draining, so tight mixtures should be avoided. If plastic pots are used, they may be slotted down the sides with narrow incisions such as saw cuts. This improves aeration and prevents root decay from over-wetness through inadvertent watering when the plants are dormant.

Development of flowers and pseudo-bulbs is best encouraged with liquid nutrients applied in weak form. It takes little fertiliser to produce good growth, but the prevailing need is for regularity rather than haphazard applications. It is frequently wondered why solid fertilisers or slow-decay fertilisers are not recommended, but the single reason could be put down as unwanted residues of fertiliser.

Plant decay or disease is mostly caused by too much humidity, often associated with high temperatures.

While it would be a worthy ambition to assemble a collection of all the cattleya species, it is not the role these plants usually assume; they are usually a subsidiary portion of the cultivated genera. Species cattleyas should never be considered the inferior part, subject to comparison with hybrids grown for exhibition purposes. The rarity and scarce supply of species orchids place them at a premium, and this applies particularly to cattleyas.

It is at times difficult to verify labels, and varietal names are often in error. It is not possible to duplicate varieties by self-pollination: much of the subsequent flowering stock of this type of propagation will carry different labellum patterns and shape and colour. Cross-pollination of two varieties of the same species will give a cross-section of flowers resembling the particular species used, but cross-pollinating from one species to another will result only in hybridising and quite different flowers. It is possible that through mis-identification this has happened to an unknown extent.

The species listed below are from the northern and eastern part of South America and only *Cattleya maxima* (Lindley) comes from Peru. It is similar to some forms of *Cattleya labiata* (Lindley). The *labiata* type is well spread over the area denoted and the locality of each is indicated in one of the maps included in the original volumes of *Veitch's Manual of Cultivated Orchids*. Since those books were produced a few other localities have been discovered, but the principal habitats of all the species or, as some regard them, the varieties of *Cattleya labiata* (Lindley) were known by the beginning of the twentieth century. The states indicated by Alex Hawkes in his *Encyclopaedia of Cultivated Orchids* as the origin of some of the species are inaccurate according to other authorities.

The unifoliate species and their approximate flowering seasons have among their numbers:

C. *dowiana* var. *aurea* (Lindley) Veitch, flowering summer-autumn during the growth process.
C. *lawrenceana* (Reichenbach f.), late summer-autumn.
C. *maxima* (Lindley), spring. (Not a C. *labiata* inclusion.)
C. *mendelii* (Backhouse), spring, early summer.
C. *mossiae* (Hooker), spring to early summer.
C. *percivaliana* (Reichenbach f.), late winter to summer.
C. *gaskelliana* (Sander), autumn, but variable.
C. *lueddemanniana* (Reichenbach f.), late summer to autumn, but variable.
C. *trianaei* (Linden and Reichenbach f.), spring to summer, variable.
C. *warneri* (T. Moore) Veitch, winter to spring.
C. *warscewiczii* (Reichenbach f.), autumn.

The seasons given are those which can be traced through lists of flowering plants, and it is noted in studying those lists that a large variation is common.

The bifoliate cattleyas are a considerable group of species which are native to some of the Caribbean islands as well as the American mainland states. Some species with their flowering times are:

C. *loddigesii* (Lindley), autumn. It is variable because of origin and the variety *harrisoniae* or *harrisoniana* may flower as late as spring. This is the only bifoliate which could be termed cool-growing, and it should be included in all species cattleya collections.
C. *intermedia* (Graham), spring. Morphologically similar to C. *loddigesii* and frequently mistaken for it.
C. *skinneri* (Bateman), spring.
C. *acklandiae* (Lindley), autumn.
C. *bicolor* (Lindley), autumn.
C. *granulosa* (Lindley), autumn to winter.
C. *dormaniana* (Reichenbach f.), autumn.
C. *guttata* (Reichenbach f.), autumn.
C. *violacea* (Humboldt, Bonpland, Kunth, Rolfe), late winter, spring.
C. *bowringiana* (Veitch), autumn.
C. *amethystoglossa* (Linden and Reichenbach f.), spring to summer.
Cattleya walkeriana (Gardner) is peculiarly individual because it flowers from the rhizome and subsequently develops the pseudo-bulb from that shoot. Usually spring flowering.

Availability of species of all kinds of orchids is impossible to forecast and it is only through private sales, sales benches at society meetings or through nurseries that plants may occasionally be bought. The process may be slow, but the final result depends on persevering and not being disheartened if importation and quarantine take their toll. Do not order one plant, but rather two or even three, so that if one is lost there will be another on the bench. Having a number of plants will improve cross-pollination. Buy seedlings where and when available.

Veitch, in his *Manual of Cultivated Orchids* makes the following notes:

> *Cattleya mossiae* (Hooker), *C. mendelii* (Backhouse), *C. percivaliana* (Reichenbach f.), and *C. trianaei* (Linden and Reichenbach f.) all flower from mature growths. *Cattleya gaskelliana* (Sanders), *C. dowiana* var. *aurea*, *C. lueddemanniana* (Reichenbach f.) *C. warscewiczii* (Reichenbach f.) and *C. warneri* (T. Moore) flower from the new growth. *Cattleya labiata* (Lindley) usually flowers after the growth has been completed and rested for a few weeks. The sheath should remain green during this period.
>
> If for some reason the growth is incomplete by the time the normal flowering period comes around, the plant may miss altogether. This could occur because of poor environment or lack of fertiliser or lack of air flow about the plants. Incorrect environment could alter the timing of the flowering by a matter of some weeks. The plant either misses in that year or carries the growth process forward to erratic habits and lack of flowers for perhaps 18 months.

For those who collect names as well as orchids, the colloquial native name for *Cattleya leopoldii* (Verschaff ex Lem.) is *Ocztictepesacuxochitl*. Some authorities consider this species a variety of *Cattleya guttata* (Lindley).

As one of the unfortunate results of our cultural methods, cattleyas appear to produce their best flowers when the plants reach or overhang the edges of containers. It should be realised that the natural tendency of all these plants is to consolidate and increase their growth, and confinement to a container stultifies this. When the plant has produced a certain number of pseudo-bulbs, a maturing process occurs. The environment and the size and nature of the place the plant occupies are the only limits on its correct and ultimate size.

In artificial cultivation, plants should not be allowed to overgrow their containers. Plants should be flowered on the next new growth. As soon as roots show or the older ones develop new tips, the plants should be repotted. This, of course, halts the plant in its growing process. It is possible to negate this loss of flowers for two years, and perhaps more, by growing two plants of each desirable clone, one slightly lagging behind the other. In this way a good system prevails and there are fewer untidy-looking plants to deal with when repotting comes around. The temptation to overgrow, for all its supposed rewards in flowers, should be resisted, particularly with species cattleyas.

Ceratostylis (Blume) 1825

A medium to large genus of orchids of the western Pacific region, stretching from the Philippines south to Niugini, where it reaches its greatest numbers. There are twelve or more species endemic to the Philippine islands.

As a genus it extends to the Himalayan region of India with a lesser concentration of species through to Malaysia. The species do not command much attention from either collectors or growers, with the result that habitats have not been denuded to the degree that applies to other genera.

Although these habitats vary in elevation from 300 to 500 metres (1000 to 1650 feet) for some species, with others higher again, the genus is not recommended for cool cultivation and should be grown in intermediate climates and shade which is not too dense. Rudolf Schlechter found some species in Niugini above 1000 metres (3250 feet).

The morphology of the genus varies from species to species, with some having deeply channelled terete foliage and others with a leaf above a short peduncle. The plants usually have a pendulous habit, which makes them suitable for slab mounts of treefern, cork, or pieces of timber which have been well weathered before use.

Nearly all species originating in Niugini have flowers which are typical of many other genera from that island, in that they are almost closed when fully developed.

As a genus there is little information other than botanical description available. The species most commonly cultivated appears to be the one illustrated, *Ceratostylis rubra* (Ames), from the Philippines. This species has fully opened flowers when developed and semi-terete leaves which are channelled. Flowers of many species develop in the axil of the leaf and are borne on short stems.

Helen Valmayor describes the Philippines species fully in her books *Orchidiana Philippiniana* and features *Ceratostylis rubra* (Ames) in the volume containing the illustrations.

It is possible that such species as this would be worth the effort of obtaining and growing as botanical subjects. But a collection, although it would offer unusual features, would be beyond the desires of most enthusiasts.

The isolated plants noted in flower, usually in spring, appear to be additions to existing collections of species and sometimes hybrids of other genera and growing in intermediate to warm environments.

Chiloschista (Lindley) 1832

This diminutive orchid is introduced to the species range because it represents a group of almost leafless plants in which the functions usually performed by the leaves are taken over by the roots. This genus is related to sarcochilus, although its origins are Indo-Asian.

The plants are epiphytic and stretch the conception of orchids into the limits portrayed by some Central and South American genera such as catasetums. The question such orchids pose is how they live without the benefit of the processes performed by leaves. The root system is frequently a tangled mass and the inflorescence arises from what could be termed a 'crown' at the base where all the roots have a joint existence.

There are few species in the genus chiloschista and they grow in moderate warmth in what is usually termed an intermediate environment. While there is no indication of the state of growth of the system as there is in leafed genera, it must be judged as well as may be by the appearance of growing root tips on older roots or of new roots emerging from the common base.

As these plants are small, fertilisers should be applied sparingly to the open root system; otherwise the plant (if such an orchid could be called a plant) will quickly lose vitality and shrivelling of the roots will indicate that it is dying.

This orchid is suitable only for slab mounting. A hard mount such as a section of a coarse-barked tree like an oak is best for tying the root system to, preferably with strips of old hosiery rather than cutting materials like fishing line or wire.

Australia also has a member of the almost leafless orchids in *Chiloschista phyllorhiza* (F. Mueller). It is frequently listed as totally leafless but may also bear small degraded leaf-like processes at the apex of the root system.

There are leafless Australian epiphytes which have the same growth habit as chiloschista — a small group named *Taeniophyllum* (Blume) comprising, as presently known, seven species. These orchids are very difficult to locate but, as with most epiphytic orchids, the roots are apparent as they creep over bark surfaces. The roots will tell you little about the plants; it is only when they produce their minute flowers that they may be identified.

All these orchids are curiosities grown for their botanical interest rather than showiness.

Chondrorhynca (Lindley) 1846
Cochleanthes (Rafinesque) 1836
Warscewiczella (Reichenbach f.) 1852

It could be asked how an orchid could come by such a name. The name refers to part of the flower, the rostellum, and some may say that we suffer for it. The genus is referred to in the *Annals of the Missouri Botanic Gardens* as 'tufted epiphytic herbs without pseudo-bulbs'. That is the important part of the plant description. The lack of pseudo-bulbs signifies lack of storage capacity to tide plants over drought or deprivation. Terrestrial orchids have the characteristic more frequently than epiphytes.

The genus is confined to the Central American states, some of the Caribbean islands and parts of South America. It is not often cultivated and is usually regarded as suited for sub-tropical areas rather than tropical, with some species from higher altitudes suitable for intermediate conditions. These temperatures and other requirements appear in the section on culture at the beginning of the book.

As the nature of chondrorhyncas is different from pseudo-bulbed species of many genera, so is their general care. Only at certain periods in growth is it wise to mist overhead or spray the plants. This means, of course, that they become individual responsibilities and that in turn can mean neglect. Their growth habit is also different, with proliferation from leafed offsets from the base of mature growths.

As the genus has few members the species are not readily available. If purchased and put through some of the quarantine procedures adopted by various agricultural authorities there could be total loss if the foliage reacts to the sprays and dips used. Seedlings are the better way to obtain plants.

Flowering is sparse on all the species, usually solitary and rather soft-looking. They should not be sprayed or misted, but the plants should never be allowed to dry out completely for longer than a day or two. That most of the species originate in humid forest regions gives some indication of the environment in which they grow. While it should not be thought that there are no plant losses in nature where only the fittest survive, in artificial conditions where there *is* control plants should not be lost.

Flowering period would vary considerably over the warmer period of the year from grower to grower. Lasting quality is short and controlled to a degree by the relative conditions in which the plants grow.

There are a number of synonyms or superseded generic and species titles, two of which are included above.

Cirrhaea (Lindley) 1832

This small genus, totally Brazilian and allied to the gongoras, was first brought to the notice of species cultivators by Conrad Loddiges, a celebrated plant culturist of the early years of the nineteenth century. In 1823 he flowered a plant of *Cirrhaea viridipurpurea*, now known as *Cirrhaea dependens* (Reichenbach f.), (see page 122). There may be some doubt about this plant, but it was imported into Australia as such about 1960. Hooker was the author of the name *C. viridipurpurea*, but his title was incorrect. It appeared in Hooker's *Exotic Flora*, which was published in 1823, where he named it as a gongora.

The cirrhaeas are epiphytes, very rarely lithophytes. They resemble bifrenarias, with the pseudo-bulbs ribbed and never fattening out to fullness in cultivation. The foliage of *C. dependens* (Reichenbach f.) is somewhat longer than bifrenaria foliage and is cast as the pseudo-bulbs age. The flowers are not bifrenaria-like.

The genus, which contains about six species according to some authors, is confined almost solely to southern Brazil and the habitats are in what is often called temperate environments, but which properly should be considered as lightly tropical, with a very adequate rainfall. In their habitats the plants grow on bare branches and trunks of trees in fairly bright light.

In cultivation the species may be grown as suspended plants, encouraging the inflorescence in its natural habit of falling away from the plant, rather than tied erect to a stake. The potting material should be bark based and free draining, considering the nature of the root system. Alternatively the plants may be mounted on slabs which are suitable for root attachment; any browning off at the tips of roots indicates that the mount is unsuitable. Slabs of treefern may prove unsuitable, but they are worth trying.

It is difficult to find illustrations of the genus in most books on the Brazilian orchids, but one in particular shows the morphology of some of the species. This is Hoehne's *Flora Brasilica Orchidaceas*, which unfortunately has not been translated into English except for a few excerpts. The illustration is unverifiable.

As with other genera described in this book, cirrhaeas may be hard to find in plant lists and probably the only source would be import and quarantine.

Comparettia (Poeppig and Endlicher) 1835

A small genus of about ten species native to Brazil, Venezuela, Colombia, Peru and extending into Central America and some of the Caribbean islands.

The most widespread is *Comparettia falcata* (Poeppig and Endlicher), which is natural to nearly all the countries over which the genus is spread. The largest of the species is *Comparettia macroplectron* (Reichenbach f. and Triana), native to Colombia, shown on page 124.

The species in general may best be described as a bundle of leaves and roots, because the pseudo-bulbs are very small and insubstantial. The plants are epiphytes and the root system in good cultivation should follow this pattern on sections of branch cut from suitable timbers. These branch sections should be chosen carefully, they need not have rough bark, but should be from deciduous trees such as oaks or similar hard-barked specimens.

The root systems of comparettias are their vulnerable points and the plants generally should be carefully watched for root activity. There is a definite growing period and it will occur in the warmer months of the year, with flowers appearing in autumn and winter on some species. *C. macroplectron* usually flowers in spring to summer.

Plants are sometimes obtainable through suppliers who import Central and South American orchids. They do not stand up well to common misunderstandings by those handling them after pre-release treatment, principally because of their delicate morphology and the quarantining methods.

It is perhaps appropriate to quote the culture given the genus about 1885 to 1890 as outlined in *The Orchid Album*:

> This plant [*Comparettia macroplectron*] will thrive in a basket or pan suspended from the roof, where it can receive a great amount of light, and in order to grow it successfully this plan should be followed. The species, which are all small growing subjects, are found on the branches of trees where they obtain just sufficient shade to screen them from the burning sun, and where they also get heavy dews at night, and sometimes a great amount of rain. It is during this rainy period that they make their growth and produce their flower spikes. Both the foliage and bulbs being small, they require careful attention as regards moisture at the roots.

In the intervening years nothing much has changed except potting material. Provided they are not over-watered, orchids grown in this way should be no more a problem than other genera. It is easy to find hanging space for a plant as beautiful as *Comparettia macroplectron*.

Constantia (Joao Barbosa Rodrigues) 1877

This small orchid is representative of a group of Brazilian orchids existing in a climate which has obviously fashioned them through evolution to withstand its rigours. The genus was named in honour of the wife of Rodrigues, who was his constant companion and assistant in his work on the Brazilian flora. Rodrigues was of Portuguese descent and unfortunately his work has seldom been translated into English.

There are three species in the genus, two native to the southern areas of Brazil in Guanabara and Santa Catharina, with the illustrated species *Constantia cipoensis* (Porto and Brade) native to the harsher regions of Minas Geraes, to the north of Rio de Janeiro.

The climate of much of Minas Geraes matches the land in its harshness and it is described in some of the literature as varying between day and night with a maximum of 30 degrees (86°F.) and more in the daytime, frequently falling to zero (32°F.) at night.

The habitat in which *Constantia cipoensis* exists is largely open savannah at medium elevations; the soil and sub-strata are such that few trees are present except along streams. The host of this orchid is principally a form of villozia and as will be noted from the illustration on page 125 the pseudo-bulbs closely attach to the rough bark-like surface left by the deciduous fronds or leaves.

Regard for such habitats and any trees which carry epiphytes brings into focus the part played by hosts and the attraction of particular hosts to different orchids.

It is extremely difficult to grow such orchids in containers: they usually die when removed from their natural growing surfaces. This specimen of *Constantia cipoensis* was imported still attached to its host and went through quarantine with the requisite methyl-bromide gas treatment. This treatment is harsh enough to kill many orchids and badly affects others.

The other two species of the Brazilian constantias, *C. australis* (Cogniaux) Porto and Brade and *C. rupestris* (Barbosa Rodrigues) grow in quite different environments, but little information is available except about their preferred climate. *C. rupestris* grows in the climate in which some 60 per cent of the Brazilian genera and species are found. This is a somewhat softer type of climate than that of *C. cipoensis*, with regular changes in temperature and humidity.

None of the genus would be recommended for either cold or warm conditions, but *C. cipoensis* thrives in moderate environment with a minimum of 10 degrees (50°F.).

Water indiscriminately applied quickly kills off flower buds.

Cryptochilus (Wallich) 1822

The true extent of proliferation of the genus and the number of species comprising it are unresolved questions. The literature on them is sparse, yet the genus has been known for such a long period that more could have been expected. In *Sanders' Orchid Guide* from the earliest years of the 20th century two species were recorded and little more has emerged in later literature.

The genus is allied to eria, but comparison of that genus with cryptochilus leaves little information other than this fact: two species are recognised from the Indo-Asian area — *Cryptochilus lutea*, which has the synonym *Eria meirax*, and *Cryptochilus sanguinea* (Wallich). However, the latter species periodically appears in collections, perhaps because of the flower colour. Red in any orchid is an attraction and it is the predominant colour of *C. sanguinea*.

Morphologically cryptochilus differ from eria in that they develop pseudo-bulbs unlike those of eria and the flower stem emanates from the base of the undeveloped pseudo-bulb. Subsequently this bulb matures, but never to the level of such orchids as coelogynes or other bulbed genera. Eria plants frequently resemble dendrobiums.

However small the individual flowers of *Cryptochilus sanguinea*, when they appear on a spike about 15 to 20 centimetres high (about six to eight inches) as a red inflorescence the overall effect is striking. If the plants are well grown, it is possible to have more than single spikes and this adds a spectacular orchid to the group of plants grown in intermediate temperature environments.

The species grow in close association in northern India and into the Khassia Hills, frequently in conglomerations of detritus which provides the food source and anchorage for them.

The plants are almost terrestrial in habit, but damp and sour conditions should be avoided and well-drained material used as a potting base. It is possible that ordinary open-type cymbidium mix would suit them, but it should be of a nature to stand dehydration in what would be inactive stages of the annual cycle.

It is always unwise to predict flowering periods of any orchids. Cryptochilus would need to reach almost a half-grown stage for the flower spikes to appear and some care should be taken with moisture control and watering of the plants when this half-grown stage appears to be reached. Water lodging in the outer and lower bracts should be avoided. Fertilisers should be moderate and not too richly compounded.

Cuitlauzina (La Llave and Lexarza) 1825

The above authors stated: 'We have named this in memory of Cuitlahuatzin, Governor of Iztapalapa, who, with much toil, has taken pains to embellish the gardens of western Mexico with the rarest of trees and plants.'

The name has been resurrected in recent times, and applied to plants once known as odontoglossum, namely *C. pendula* (La Llave and Lexarza), which has the synonym also of *Odontoglossum citrosmum*. The present name of cuitlauzina appears in *Generic Names of Orchids*, by Schultes and Pease and the spelling is taken from that book. Leon A. Wiard, in *An Introduction to The Orchids of Mexico*, lists only this species.

Cuitlauzina pendula is not hard to grow if it is given moderate warmth, rather than the cold conditions in which it is supposed to thrive. It is possibly best listed as cool-growing, where 'cool' is understood to mean temperatures of not less than 10 degrees (50°F.). Shade should be about 75 per cent.

The plant form or morphology of this species would lead one to regard it as a member of the oncidium group, and the first plant I bought was sold as such. Once known, however, such an error could scarcely occur a second time. It flowers in spring.

There is something to be learned about growing *C. pendula* and, as with most orchids a little study soon unravels the mystery. Unlike an odontoglossum, it must be grown in conditions where there are periods of dryness when the plant may subsist almost on the natural humidity of a glasshouse provided it is not less than 50 per cent. However, once the flowers begin to open this humidity will quickly spot them, and a year's work could be spoiled in one night.

The flower spike itself, once it shows in the very immature growth sheaths and as the first tip emerges from them, must be kept free of moisture at all times or it will quickly decay and again there is a year's work lost.

The plant should be fertilised like all orchids with liquid fertiliser once it commences to grow, with care taken not to overwater or over-fertilise. It should preferably be grown as a suspended plant potted in coarse bark with a little coarse gravel added. Once the flower spike is gone, the plant should be kept slightly shrivelled until new growth again shows in early winter.

Maturing of the pseudo-bulbs is very slow after flowering.

Cynorkis (Thouars) 1822; also *Cynorchis*

This group of orchids, in most publications listed as cynorkis, is based principally on the islands east of Africa, but several species are found on the African continent. The genus is related to habenarias, is mostly terrestrial and appears to be seldom cultivated, despite the obvious attractions of the species.

Terrestrial orchids are less popular among cultivators than lithophytes or epiphytes, perhaps because of the difficulty in getting together a suitable potting mixture. The mixtures used for such semi-terrestrial orchids as cymbidiums are too dry to suit orchids which have an underground tuber system.

There are more than one hundred cynorkis species and they vary considerably in morphological terms. Unlike most orchids, the foliage is impermanent and renewed each year.

Thouars, the man who named the genus, was associated with the islands where this genus is based. Madagascar has a large orchid population, but little information is available about it and possibly it is better known as the origin of the angraecums. Thouars had associations with La Perouse, who disappeared after he left Botany Bay in 1788, and despite a search was never found.

The species illustrated is typically terrestrial, with leaves similar to those of a tulip: the large funnel-shaped leaf and its associated smaller leaf enfold the flower spike. The plants vary considerably, and *Cynorkis ridleyi* is of medium size.

The problem of growing these deciduous terrestrial orchids in a general collection is that they must be isolated once the foliage starts to decline. Some species grow in moist situations, others in dryer areas. They are not always completely deciduous. In taking them into cultivation growers should understand them and place them where the tubers will not become too moist in dormancy and decay. Each species is individual and should be treated so.

The growing medium outlined for disas should be tried, and modified if it proves unsuitable. Cultivation generally parallels that of disas except for the temperature range. For cynorkis temperatures should be intermediate to warm, depending on origin.

Shade for most terrestrial species is light and should be tried at the density used for the normal orchid collection which sustains good ranges of sunlight through the year, and is diffused with weaves of about 50 to 75 per cent, depending on its intensity.

Cypripedium (Linnaeus) 1753

A genus which, as the date of its naming indicates, has been known for more than two centuries. During this time it has, after occupying suitable areas of the northern hemisphere, been so completely destroyed that it has disappeared as a natural growing plant in many countries. It is said, with whatever truth, that only a very few plants of *Cypripedium calceolus* (Linnaeus) are known to have survived in Britain where it was once recognised as a rather common wild flower in woodlands.

It is difficult to give a figure for the number of species, but several survive and are frequently photographed and remarked on in the United States of America. Very few countries of the northern hemisphere are without representatives of the genus, the most widespread of which appeared to be *C. calceolus*. The plant photographed (see page 127) was exhibited in an orchid show in New Zealand and represented there a genus difficult to cultivate, yet in the right environments, unbelievably simple. The species *C. japonicum* (Thunberg) is native to Japan and China and possibly also suffers the same rate of eradication which affected the populations of the plants in other countries. Living pressure in most instances is the cause of its disappearance, with the felling of timbered areas destroying its environments faster than any other factor.

Importation and subsequent reconditioning of such plants after quarantine procedures may make the project almost impossible. There is no pseudo-bulb system to reconstitute leaf destruction and as this leafage is susceptible to damage because of its size and lack of substance, the plant does not easily withstand incorrect dispatch time, poor handling and long periods in close confinement, let alone the added impact of chemical decontamination.

The genus as a whole is terrestrial and should be grown as such if imported plants or flasks of seedlings are available. Other than direct importation of seed these are the only sources open to growers in countries of the southern hemisphere where the genus is not endemic.

It is little trouble devising potting material for such plants, but there may be some finer points to be found in various environments which would improve culture. With plants where there is no pseudo-bulb system there is probably no second chance if mistakes are made.

Shading should be adequate, considering their habitats in wooded areas. While too dense shade should be avoided, it would need to be somewhere in the vicinity of 60 to 75 per cent for most of the year.

Few orchids should be wet all the time, but constant moist state of the potting material would appear to be necessary and very weak fertilisers applied periodically. Slow-release fertilisers of any sort would be inadvisable, but some growers prefer them.

Dendrobium (Swartz) 1799

One of the largest genera of orchids, this group is as unique to its hemisphere as the laelias and cattleyas are to theirs. In addition to its numbers it has perhaps more diversity than the cattleyas and laelias, usually referred to as the laelieae. The dendrobieae grow in an arc commencing in Indo-Asia and extending in a wide band through the Indonesian islands, the Philippine islands and ending with a single species in New Zealand. As a genus they cover almost half the world.

Over this region they still conform in cultivation to the three general temperature ranges in which they are suited. Some are cool-growing, some will tolerate cold but are not suited to it. Others prefer an intermediate environment and will not thrive below a certain minimum temperature; others again demand tropical temperatures to grow and flower satisfactorily. Cultivation of some species has proved impossible.

Besides the temperature requirements, other groupings of the genus demand special conditions for correct culture. Among these dendrobiums are what are usually termed the nigro-hirsute species. These are characterised by slight or extensive covering of what are best described as dark hairs. These extend also to the sheaths covering the flower buds and sometimes the base sections of the flowers, ovary and pedicel.

The origin of these hairs is probably climatic and designed to protect the plants from intensified moisture covering the growths and flowers. As a group they are generally found in deeper shade than other dendrobiums, subject to condensation and moisture which does not evaporate as quickly as from plants more openly situated. Another characteristic of these dendrobiums is that in general they have pale to white flowers.

The colour of orchid flowers frequently indicates how much light is needed in cultivation, with red to yellow flowers tolerating brighter conditions than white or green. Occasionally this may be noted in such dendrobiums as *D. nobile* var. *virginale* when grown with normal coloured forms, which grow in more robust fashion than the albino form. They should be separated.

This is only one of many restrictions when formulating conditions designed solely for species orchids. It could be thought that dendrobiums which come from similar altitudes and share other necessary conditions could be grown together. This mistaken notion causes failures. In very few instances are dendrobiums collectively grown as a genus and there are several reasons for this. The genus suffers neglect, principally because the species are unavailable.

As cultivated plants, although the root systems are best suited by open mounting, the plants in general conform to container cultivation. In this state they will send out some aerial roots. It is not possible to contain the roots entirely, but they are seldom as invasive as the roots of many vandaceous orchids.

The size of plants in many instances sets growers against these orchids as a main genus around which to build a collection. Unless the flowers emanate from the apex of pseudo-bulbs they must be given adequate light over the whole length of the pseudo-bulb. An example of this would be in trying to grow orchids like the soft-cane dendrobiums such as *D. nobile* with tall-growing species like *Dendrobium moschatum* (Swartz), the flowers of which are produced at the apex of the pseudo-bulbs. One is a hard-cane type, the other soft-cane and flowering, in good examples, down the length of the pseudo-bulbs. It is not a difference in environment so much as habit; an instance of poor combination, too, of hard-cane and soft-cane dendrobiums.

In some species there are other factors to watch. Consider the instance of *Dendrobium heterocarpum* (Wallich), the synonym of which is *Dendrobium aureum* (Lindley). It grows in Sri Lanka, southern and parts of northern India, Burma, Thailand, the Philippines and down to Java. In all these locations the plants grow at different altitudes and in different climates. Although there are morphological differences, it could well be imagined that a plant of this species could be bought or exchanged in the understanding that it would grow in the moderate temperate to sub-tropical zone for which it was intended. In reality the origin of the clone should also be known.

The same factors do not apply as particularly to dendrobiums purchased from or exchanged with other growers when their origins are tropical. Despite the fact that they may originate at high elevations and at times in cool to mild climates, it will be noted in cultivation that they conform to warmer conditions than imagined. In this consideration latitude also plays a part and modifies the propensity of the species concerned.

In summary, it is far better to conform to cultural programs rather than indiscriminately placing orchids in what are at best chance conditions.

The genus as a whole does not have a reputation for longevity in cultivation, perhaps more particularly when species are at issue. Hybrids have a much better history and this is borne out in a plant of soft-cane type, namely a secondary hybrid named *D.* Kobayashi, which was purchased as a seedling in 1945 and is still thriving in 1988. It is doubtful if a species could be persuaded to last as long, even supposing it endured in the respect and admiration of the owner. Another set of factors takes over here, including that of 'keeping up with the Joneses'.

One of the most difficult sections of the genus to retain is what we have described as the nigro-hirsute group. It comprises several well-known dendrobiums, among which we will quote two: *Dendrobium infundibulum* (Lindley) and *D. formosum* (Roxburgh). They are similar and it is common to see one sold as the other. While *D. infundibulum* (Lindley) is capable of cool to moderate cultivation, *D. formosum* (Roxburgh) will be found to accommodate better in mildly sub-tropical to tropical climates. Both have pure white flowers with a yellow-orange central area of the labellum. They are not albino forms of a coloured species, despite carrying a well-known colour sequence that albino species usually show.

If any of these nigro-hirsute dendrobiums are taken up as adjuncts to a collection or grown as a main section of a mixed collection, it is improbable that they will live longer than the commoner hard-cane or soft-cane types. There is some indefinable thing about the environment which appears too hard to distinguish and maintain. Success may be obtained with the strong-growing types rather than the smaller and difficult species such as *Dendrobium cruentum* (Reichenbach f.) or *D. bellatulum* (Rolfe). These are two of the most attractive members of the genus and probably among the most difficult to grow.

Culturally it appears that these nigro-hirsute dendrobiums should be kept moist about the root system yet be in a good air flow maintained throughout the year. These two matters must be attended to, and the collection of orchids be treated as seasonal, with an indefinite dormant period in individual plants through the twelve-month cycle. Choice of species may depend on one's preference for open-root culture, or growing the collection in containers. Some of the nigro-hirsute dendrobiums should be grown open-rooted, while others tolerate containers.

Some of the recommendations for growing nigro-hirsute dendrobiums may appear pinpricking, particularly in their demands for individual care while not being over-kind. Remember that we are not discussing the tolerance levels of the plants but rather the set of conditions which will keep them not only alive but thriving.

It may be well at this point to sort out some of the combinations of the genus which may be grown together in intermediate conditions, leaving out those which may grow cool to cold and those which need sub-tropical conditions. The criterion is that the plants tolerate temperatures which may go as low as 12 degrees (54°F.). It is not a definitive list, but rather a recommended group:

Dendrobium nobile (Lindley). This could be accompanied by others which grow to the same height — about 75 centimetres (2 feet). There are many varieties and colour forms of this species and the best is possibly the variety *nobilius*. Red-purple.

D. pierardii (Roxburgh). This is included here because it is the direct opposite of the first-named plant. It is pendulous and must be grown as a suspended plant. It may also grow as long as and at times have longer pseudo-bulbs than *D. nobile* (Lindley). Pale lavender coloured.

D. bensoniae (Reichenbach f.). In some books referred to also as *D. signatum* (Reichenbach f.). This orchid has a habit which could be termed untidy and many of these soft-cane dendrobiums will be found so. It is gold in colour and does not fit the original description of *D. signatum* (Reichenbach f.) used by hybridists, which is yellow.

D. chrysanthum (Wallich). There will be a temptation to tie this species erect but it should be resisted. The plant appears to flower more consistently when allowed its natural pendulous habit. It is gold-coloured with two dark red-brown blotches in the labellum and flowers in summer.

D. findlayanum (Parish and Reichenbach f.). The pseudo-bulbs are unmistakable, like a string of large flattened beads. The flowers are pale lavender and gold, sometimes almost white and gold. It also is an untidy plant needing adequate light and room to spread its pseudo-bulbs.

D. thyrsiflorum (Reichenbach f.). A hard-cane dendrobium which grows erect, the flowers gold and cream or white, with the variety *densiflorum* golden and at times bronze tinged. Like *D. chrysanthum*, these two flower later than the spring to early summer flush of the soft-cane types.

D. lituiflorum (Swartz). Similar to a pale form of *D. nobile* (Lindley), but with more open and straggly flowers, which come in late spring to early summer.

D. crepidatum (Lindley). Almost a miniature with the most delicately tinted flowers of all the genus. It is principally pink and white with crisp, glistening flowers, rather small and at times tinted with pale lavender.

The above are only a few of the many such dendrobiums available for cultivation at times in Australia. They would be unsuited to many regions, but have scope in the right environments existing in temperate to sub-tropical parts of Australia. None of these are tropical species.

There is also a hard-cane group suitable for cultivation in similar climates and it includes *Dendrobium chrysotoxum* (Lindley), *D. farmeri* (Paxton), *D. aggregatum* (Roxburgh), *D. gibsoni* (Paxton), *D. fimbriatum* (Hooker), and *D. sanderae* (Rolfe), which except for the last-named are yellow to gold, indicating fairly bright light in cultivation.

In the upper temperature range, which must also be subject to bright light conditions, is a very large group of tropical and subtropical dendrobiums based on *D. phalaenopsis* and *D. bigibbum* (Lindley). Although there is disagreement, *D. phalaenopsis* (Fitzgerald) is now considered a variety of *D. bigibbum* (Lindley).

With such dendrobiums as a base, tropical growers and even those in sub-tropical districts can choose from many species, although much of the colour of the temperate zone orchids is lost. Each grower should understand the restrictions imposed by their individual growing conditions and never be carried away by names, photographs or inducements offered by other growers to 'give it a go'.

Many of the orchids listed above are illustrated in *Growing Orchids*, *Book Three* and some photographs are repeated here.

Some dendrobiums can be judged by their appearance as suitable for mounting on slabs. Potted dendrobiums all grow in bark mixes or plain bark. In this there is a rule to follow: although some barks seem expensive, in general the more expensive material gives the results hoped for.

For those who prefer low-growing miniature and semi-miniature species, there is a good range from which to select. Some are pendulous, others have that habit but may be grown erect if tied that way; but most are compact plants which flower from the apex of the pseudo-bulbs or from the uppermost nodes. They may not be grown closely together as paphiopedilums will grow and should be given as much space between the plants as each occupies.

They flower mostly in spring to early summer, but some are variable in their seasons and may flower in midsummer. In some of the summer-flowering species, the life expectancy of the inflorescence is short unless the plants are kept cool. Following are some of the smaller plants:

Dendrobium crepidatum (Lindley), already described and not grown as frequently as it merits.

D. *arachnites* (Reichenbach f.), a very unusual Burmese dendrobium, giving a prevailing impression of a red to orange flower. The pseudo-bulbs are very short, 80 to 100 millimetres (3 to 4 inches) tall.

D. *capillipes* (Reichenbach f.), another very small dendrobium with stout pseudo-bulbs topped by three or four leaves, and usually producing two or three flowers on short stems from near the apex of the growths. It is a spring-flowering species which must be staged through a partial dormancy over the cooler months of the year. The flowers are white to golden.

D. *chrysocrepis* (Parish and Reichenbach f.). Grows to about 20 centimetres (6 inches), is deciduous and flowers on bare pseudo-bulbs with golden-yellow blossoms, sparsely produced.

D. *parishii* (Reichenbach f.). Short pseudo-bulbs which produce delicately scented flowers, pink to deep reddish-purple, in spring and usually from bare pseudo-bulbs from which the leaves are cast as they mature. Again a plant which must be carefully grown through winter and into flowering.

D. *senile* (Parish and Reichenbach f.). A most unusual dendrobium with the leaves and stems or pseudo-bulbs covered with short white hairs. See page 131.

D. *trigonopus* (Reichenbach f.). See page 132.

D. *crystallinum* (Reichenbach f.). Illustrated in *Growing Orchids, Book Three*. One of the most beautiful of the genus, coloured similarly to D. *wardianum* (Warner), which is also illustrated in *Growing Orchids, Book Three*.

The list could be extended over many pages for normal and miniature species and, in fact, could take the whole of this volume. If the genus is the subject of a collection it will be found that it contains many orchids which must be grown as individuals rather than treated as a mass.

To reduce culture to a formula, preferably do not mix what are recognised as 'soft-cane' species with those regarded as 'hard-cane'.

Dendrochilum (Blume) 1825

In company with other spectacular orchids grown when the cult was in its middle years at the end of the nineteenth century, this genus also produced comparatively rare specimens such as a plant exhibited by Sanders at a meeting of the Horticultural Society. *Platyclinis filiformis* (Bentham) as the species was then known, carried 120 spikes of flower. It was not a rarity in groups of orchids then but it is doubtful if such plants can be grown a century later.

The genus is native to the islands of the Indonesian archipelago, the Philippines and down to Niugini. It is principally epiphytic, occasionally lithophytic and never terrestrial. Not all the species carry the long racemes containing more than a hundred small flowers, and some have relatively few and on short stems. There are more than 140 species, none of which are cool-growing. Intermediate climates suit it.

As epiphytes the plants should be grown in containers, either plastic or clay. The potting material must be relatively coarse and never bonded into a tight mix with finer material. Drainage should be complete, so that the material never becomes wet or soggy.

One of the presidents of the Horticultural Society in Britain who was distinguished for his service to orchid culture at the end of the nineteenth century had a plant of platyclinis (dendrochilum) three feet across and carrying more than 100 spikes of flower. The owner, Sir Trevor Lawrence, had a collection which was as notable as any recorded in the history of the cult. He had several glasshouses, and moved the plants about constantly in the various houses to emulate the changes they went through in their annual cycle. His plant of *Dendrochilum (Platyclinis) glumacea* (Lindley) was accommodated in what was known as the East India house as it came into growth. As the growths proceeded, the flower spikes developing in the leaf whorl were kept dry but the plants were freely watered. Much time was spent in cultural procedures in those days. The flower spikes are not long lasting and as they ended the plant was repotted in successively larger pots. The growth is close set and room was given for only two years' additional growth. Potting material was wisdom exemplified: three parts of the container was crocks with a thin layer of sphagnum moss on top, then a mix of equal parts of good fibrous peat and clean picked moss, with a liberal sprinkling of crock material through it. The plant was watered only enough to keep the material moist. Once the bulbs had matured it was moved to the cattleya house and rested. For the amateur grower who looks at the plants only at weekends, the lesson for this genus is clear. It needs time and concentration.

Disa (Bergius) 1767

Some say 'dysa', others 'deesa'. The pronunciation is immaterial. The genus is among the largest of the South African flora of the orchid section but few are cultivated successfully in general species collections. Probably the best known and most widely cultivated of the whole family is *Disa uniflora* (Bergius) principally owing to its beauty, which places it among the most beautiful of the world's orchids.

If a recommendation could be made regarding the genus it would be that the grower discard the rest of the collection and concentrate on disas. They are amongst those orchids, such as phalaenopsis, where everything becomes secondary to their culture. Only rarely is it possible to cultivate one or more disas to their ultimate perfection when they are grown as adjuncts to collections of species from all the countries of the world.

The best species is undoubtedly *D. uniflora* (Bergius); the name is completely wrong, because it bears single flowers only when poorly cultivated or ill-suited to the environment in which it is placed. One frequently reads that disas are swamp orchids or that they all grow in wet places, but this is a complete misstatement: *D. uniflora* has been found in a number of environments but all are principally based on moisture freely available to the root system throughout the annual cycle.

The natural multiplication of the original seedling, if the clone begins as such, is for the rhizome to produce outer shoots from the base of the plant, on the tips of which new growth is formed and makes its way to the surface to form on offset. Some refer to these shoots of the rhizome as stolons. The system is typified by the Australian terrestrial pterostylis, and the disas are terrestrial orchids. The original flowering growth gives way to the new shoots, the plant multiplies and such a system should be aimed at in cultivation.

Most of the disas may be grown in sphagnum moss, but it is a rather daunting system if a total collection of disas is grown. The moss must be replaced after, at most, the third growing season. If it is left longer, the plant is subject to root rot and finally loss of the upper parts of the growing shoot.

An ideal potting system must be worked out to give perhaps three years free from worry about what the roots are up to. Use equal parts of pearlite and medium fine bark and coarse sandy gravel, with all the fine sand sieved out. A little peat-moss may be added, but only as much as one-eighth of the total mix.

A potting mix like this was used by the owner of the plants shown on pages 133 and 134 and used most successfully. Temperatures must be kept below 25 degrees (77°F.) in the warmest parts of the year; cold does not cause problems provided the plants are watched carefully, watered infrequently but kept damp and not subject to temperatures below 5 or 6 degrees (40–43°F.).

Other members of the genus include *Disa tripetaloides* (Linnaeus f.) N. E. Brown, which was used in hybridising as one of the few large flowered species; *D. nervosa* (Lindley), *D. crassicornis* (Lindley) and *D. triloba* (Lindley), to mention only a few from a long list. Most are more of botanical interest and quite an amount of variation in conditions and environment are necessary to grow only a few of the total number.

It was only through the introduction of seed and flasks of seedlings that the genus was brought to Australia and the same restrictions would apply to other countries because of the hard line taken by the authorities in South Africa about export of plants of the genus. Occasionally seed from species is offered to accredited specialists.

Diseases of fungoid type appear to be the principal detriment to plants in cultivation. As they are terrestrial orchids grown in conditions which suit the proliferation of this type of infestation, careful selection of potting material, cleanliness of equipment and isolation from infective sources is advisable. Potting material may be sterilised, but probably natural reaction between the various constituents of the environment will take care of other pests and diseases. Slugs and snails must be rigorously excluded or dealt with by baiting; in this matter prevention is better than cure.

Accommodating the plants needs careful planning and if it is intended to grow a few plants as adjuncts to collections they can be grown in the same conditions given to masdevallias in constant moisture, fresh air freely admitted and attention to the upper temperature limits.

In many years of orchid growing I never was prepared to sacrifice other orchids to accommodate disas and this information is based mostly on others' experience. A very good book which includes the genus is *Wild Orchids of Southern Africa*, by Joyce Stewart, H. P. Linder, E. A. Schelpe and A. V. Hall. This book contains a list of the species and the relative keys to them. It also describes the colour of the flowers and many are illustrated in colour.

Other books on the South African genera, while not numerous, give adequate descriptions of the principal species, but there is little information available about the source of seed or flasks of plantlets. If and when they may be bought it will be found that in correct climates and environments the seedlings grow fast, unlike other orchids which may take years to make progress and flower.

The genus is apt to be judged for colour by illustrations of the large flowers of the brilliant and variable reds and stripes of *Disa uniflora*. It is a mistake common to all orchid growing, particularly when ordering species plants, to have preconceived ideas about their colour. The disas are usually bought unflowered and a flask of seedlings of *D. uniflora* could contain a remarkable assortment, depending on the parents used in cross-pollination.

Information should be sought from orchid societies in South Africa or from growers of the genus in the country where one resides. Obviously they are orchids which are unsuited for tropical, sub-tropical and sometimes temperate climates. It should be remembered that orchids, particularly species, should not be subject to experiment.

Disas flower in early summer to autumn and appear to be quite variable according to the cultivator and the region where grown. Fertilisers should preferably be weak and applied frequently while the plants are growing and flowering. The growth process should not be prolonged and if this occurs it should be apparent that the environment is incorrect and some changes made in either temperature or air flow or shading. Shade should be judged according to the region and could be as high as 75 per cent. Leaf colour should be light green, glistening and fresh-looking; a tendency to brown tips should be analysed and the watering program corrected if it occurs.

The recommendation to discard other genera in order to cultivate disas may seem a strange one, but success with disas in a mixed collection is difficult to obtain. Disa species offer a fair variety: hybrids are easily obtainable and perhaps as many as ten or more species. Altogether such a project in the right environment will give as much enjoyment as a collection of mixed genera.

I grow a few disas in 70-millimetre plastic pots packed lightly in sphagnum moss. These small pots were stood on a layer of gravel about five millimetres deep (¼ inch) in the bottom of two-litre ice-cream containers.

Just below the level of the gravel, but not in the bottom of these containers, I punched small holes in the sides of the containers to allow excess water to drain away.

These disa offsets were not watered directly nor the moss dampened except for what capillary action allowed by the moss. After the hot summer and the chill of winter has begun, the plants grow in almost total outdoor shelter of a plastic roof with soft-cane dendrobiums and flower in their season. The temperature over this period fluctuated from 35 degrees (90° F.) to as low as 8 degrees (45° F.).

The gravel was not allowed to dry out at any time and the sphagnum moss remained healthy but not growing. The gravel was always watered with a weak nutrient solution.

Doritis (Lindley) 1833

This genus was first known by relatively small scraps of plants and it had a chequered history during which it was known as a phalaenopsis, although the original ticket was inscribed by the finder, Finlayson, as 'Epidendrum, Turon Bay, n. 521'. Reichenbach in 1874 described an obviously similar plant as *Phalaenopsis esmeralda* from Indo-China. A second species has been credited to the genus in *Doritis regnierana*, also described by Reichenbach. Not all taxonomists agree with this.

Similar species were added to and subtracted from the genus at times, but in effect only one species is now apparently recognised in most publications, although Haruyuki Kamemoto and Rapee Sagarik in their book *Beautiful Thai Orchid Species* appear to recognise others. The taxonomy other than for *Doritis pulcherrima* (Lindley) remains confused.

There is considerable colour variation, from the depth shown on page 135 to paler almost white forms. The flower spikes are erect and may carry twenty or more blooms, which open in sequences of four or five upward to the apex. In conditions of maximum potential more flowers would be expected to be open, but the lower blooms fall as the spike and flowering lengthens. Shading should be moderate, not dense.

The genus is listed as native to the Indo-Chinese peninsula, Sumatra, and associated islands, growing almost as a terrestrial or lithophytic orchid and seldom epiphytic. In its habit of growth it is similar to phalaenopsis and is best grown in sub-tropical or tropical climates, although preferring low temperatures at periods during the annual cycle.

D. pulcherrima is a particularly hardy orchid; in its native habitat it can survive hot dry periods of four or five weeks. In these conditions it will shed all foliage from the elongating rhizome; when the climate improves it replaces the leaves and goes on to produce flowers. In cultivation the plant may react in a similar way, but any period longer than that quoted would be most unfavourable. In ideal conditions, of course, the foliage would be cast only as it aged and was replaced by new leaves, with at least four or five being retained permanently on the plant until their turn came for replacement. This would mean that the leaves were retained for up to four years before being cast. They resemble short phalaenopsis leaves.

Give liquid fertilisers in weak concentrations, and use open-textured potting material to encourage the roots to adhere to containers. Pots or slat baskets may be used. *D. pulcherrima* (Lindley) usually flowers in autumn.

Encyclia (Hooker) 1828

Although the genus was formed so early in the history of orchid growing, many species now included in it have been added only in recent times. It was common in the earliest days of the craft to include all epiphytes in the associated genus epidendrum and it is from this source that the genus encyclia members have largely been drawn.

It may be something of a problem for the uninitiated growers to know where to draw the line of distinction between the two genera. Generally, epidendrums have reed-like stems or pseudo-bulbs, whereas encyclias have distinct pseudo-bulbs topped by leaves, from the centre of which grow the flower stems. The analysis of the two forms is clearly outlined by Dressler and Pollard in their book *The Genus Encyclia in Mexico*. The other prominent distinction is in the labellum and column, the central parts of the flower. In epidendrum flowers the column is completely united with the labellum, whereas in encyclias the two are quite or partly separated. There are other points of difference, but these are usually sufficient to distinguish the two genera.

A quick examination of a flower of *Epidendrum* Obrienianum, the common red epidendrum often called the crucifix orchid, will soon give meaning to the description. The labellum or lip is the part responsible for the origin of its common name and the labellum will be found to be completely one with the column, which ends just above the two small yellow calluses.

Not all encyclias and epidendrums follow the pattern, but the majority do so and it is a good way of identifying them. Some authorities do not accept the separation of the two and in a great amount of literature it is common to find them indistinct or not separated. However, in most material published from 1960 onward the plants are correctly named. There are other points of divergence, too, with some names containing the suffix 'um' rather than the suffix 'a'. For example, *Epidendrum vitellinum* should appear as *Encyclia vitellina* (Lindley) Dressler.

Lindley was the author of the species name in 1831, based on a Mexican plant and included in the genus epidendrum. Dressler moved the species to encyclia in 1961 and altered the suffix.

The genus encyclia is common to much of Central and South America, also the Caribbean islands. Some would say that the morphology is stereotyped and the flowers similarly categorised, but more than one hundred illustrations could be produced demonstrating great variation.

Authorities such as Alex Hawkes included all the genus encyclia under the heading epidendrum and in his introduction to the genus remarked that it was one of the largest families of plants in the orchid world.

Mexico has a number of attractive encyclia species with marked variation in flower patterns. It will be remarked by observant growers, however, that while some of the flower habits conform to the dorsal sepal uppermost in their presentation, others, in similar fashion to the reed-stemmed epidendrums, have the labellum uppermost and the flowers back to back. The illustrations will indicate these habits and do so in the book *Growing Orchids, Book Two*.

The growth of encyclias generally follows a singular pattern, with new pseudo-bulbs made from extensions of rhizomes, formation of new sections with mature pseudo-bulbs, and flower from the apex of these new growths. If for some reason the flowering does not follow directly from these new growths, they will seldom flower until new growth is added in the following cycle. This would indicate that the cultural pattern was wrong, and adjustment of the environment and climate would be necessary.

While some encyclias may be grown in containers, others will prefer slab mounting, preferably on pieces of treefern rather than the harder wooden mounts sometimes used. One such would be *Encyclia polybulbon* (Swartz) Dressler, a creeping type with a longer gap between the pseudo-bulbs than on the closer-set container-grown plants. It is a most attractive encyclia, proliferating into a plant mass which quickly outgrows small pieces of treefern.

With these slab-mounted plants, a watering system which should be adopted is to soak the back of the treefern mount every time the plant is watered rather than soaking from the front. The new growth on *Encyclia polybulbon* may appears at any time through the annual cycle and is easily lost when too much moisture is applied to the plant.

For those interested in the genus, particularly relating to Mexico, the book by Dressler and Pollard is a good investment. They are neither 'splitters' nor 'lumpers', but divide the genus into several recognisable sections.

It is best to regard the genus as intermediate to cool-growing, with few species in the latter category.

Among the unusual encyclias transferred from other genera must be included *Encyclia mariae* (Ames) Hoehne and *Encyclia citrina* (Llave and Lexarza) Dressler. The first-named is not unusual in its morphology, but rather in its flower. It is illustrated. The labellum is the outstanding part of the flower, which has petals and sepals of green to lemon-yellow setting off its brilliant whiteness.

Encyclia citrina is perhaps the more unusual of the two because of its cool-growing reputation and the silvery-grey foliage. The flowers, borne singly, are a beautiful golden-yellow. Also remarkable is its pendulous habit and the tendency to grow it erect must be resisted. This form of growth, of course, means that it must be cultivated on a slab of treefern or rough-barked wood. The root system is vulnerable to dampness and the rather shrivelled appearance of the pseudo-bulbs is natural. In many years of orchid cultivation it is seldom that it can be kept in condition like other orchids in the genus, and it is a cultural and environmental headache to most growers.

Potting techniques for the genus comply with the usual rules of not over-potting the plants and keeping the material open and free from any tendency to sourness. Bark is most commonly used and it may be blended with charcoal or coarse gravel. It should be possible, also, to grow many encyclias in stones of about thumb-top size, with smaller gravel making up a quarter of the mix and small shreds and tufts of moss distributed through the material. This forms a quick-drying medium which should suit such of the genus as *Encyclia vitellina* (Lindley) Dressler. It has softer pseudo-bulbs and foliage than some others like *Encyclia cordigera* (Humboldt, Bonpland, Kunth) Dressler, which has laelia-like pseudo-bulbs and tough, stout foliage.

Each member of the genus must be handled as an individual part of any collection and fitted into the environment of the enclosure in the place which appears most suited. In general cattleya conditions would be recommended as best, except in the instance of *E. vitellina* or *E. citrina*, which some designate cold-growing. They are not so, but finding their correct niche in orchid cultivation may be frustrating.

The genus, in common with other Central and South American genera, is subject to continual revision. Some of the names given in this book may be superseded by the time it is printed. A glance at the 1985 handbook conveys some idea of the scope of these changes, particularly in epidendrums.

Encyclia polybulbon (Swartz) Lindley is a recent transferree to the genus *Dinema* (Lindley) 1831.

Epidendrum (Linnaeus) 1753

This name was among the early terms used by botanists to distinguish genera one from another. In reality it misrepresented species as belonging to this branch of the laelieae and it was left to botanists of the 19th and 20th centuries to correct these mistakes. Some species are still not transferred to their rightful designations and the work continues. A problem associated with the original errors is that so many plants are best known by their wrong names and it is significantly more difficult to separate them because of inter-generic hybridising.

To exemplify this, we can quote *Sander's List of Orchid Hybrids* relative to the genus encyclia. In the sixth five-year addition to this list for the years 1981 to 1985, thirty-six encyclias are listed simply by their species name under this genus and referred to the genus epidendrum for the named hybrids attributed to them. It is most unlikely that there will ever be a change in Sander's list because encyclias are only one of many genera transferred to their correct classification. This throws the whole system of orchid taxonomy into doubt; one can only prefer that taxonomists would leave things as they are and note by synonymy the preferred title.

Reed-stemmed epidendrums are among the most numerous of the Central and South American genera — not in numbers of species but in numbers of naturally growing plants. In truth, if they were more popular it is possible that they would have been by the end of the twentieth century an endangered species like so many others.

The nature of their growth is such that in correct climates and environments they have the capacity to spread rapidly by vegetative means, as well as by seed, allowing for the profligacy of nature in the quantity of seed cast. In home-grown plants it is common for a head of flowers to carry seed pods. Seed will not survive long unless it is scattered on a receptive medium and fungal diseases are controlled.

Epidendrum ibaguense (Humboldt, Bonpland, Kunth), syn. *E. radicans* (Reichenbach f.), one of the commonest and freest growing of the species, is suited to outdoor cultivation. In gardens in quite cool climates, provided the plants are not frosted, they are capable of flowering for a large part of the year. In the southern hemisphere, they will flower more prolifically and for longer periods when they are backed against a sunny north-facing brick wall. In such situations they will seldom suffer sunburn on the hottest of days. In some countries where these orchids grow they are part of the garden in numerous varieties and in some districts in Australia they grow simply as garden plants which must be restrained and periodically thinned-out by removing some of the growth and distributing it to other growers or discarding it.

The colour of *Epidendrum ibaguense* (HBK) varies from white through cream and yellow to red; this species and similar variants are common throughout much of the northern areas of South America and Central America.

One of the largest of these species, colloquially known as the 'crucifix orchid', is *E. cinnabarinum* (Salzmann) and it represents a group which needs intermediate temperatures to grow and flower. The minimum temperature should be no less than 15 degrees (60°F.) and preferably a little higher.

Tolerance to light in all reed-stemmed epidendrums is fairly constant, but the softer-stemmed species such as *E. stenopetalum* (Hooker) and *E. medusae* (Lindley) are self-selective and need more careful shading than the harder-looking epidendrums such as *E. radicans*.

In these reed-stemmed types, the root system may emerge at any section of the stem and the flowering head appears when the growth or stem has reached its maximum height. The root system is the wandering type and will quickly invade decaying vegetation or other sources of food. When potting these orchids, the container should always be large; the potting material should be rough and contain matter such as leaf-mould or decayed lawn clippings. Organic manures or inorganic liquid fertilisers should be applied with discretion. It is not necessary to remove the flowered tip of exhausted spikes, as these frequently produce secondary flowering spurs rather than shoots which grow from the lower or median parts of these stems.

The lists of names in older literature are, of course, out of date, but the means of referring each genus to its rightful name is described in the section on encyclias (page 64). If in doubt, it is better to leave plants as labelled when bought or otherwise obtained.

The commoner species are easily obtained and should be grown with as much enthusiasm as other orchids. The more care and attention given the better the heads of flower, which last for some months by progressively developing flowers in the central rosette of buds.

As with all monopodial orchids, epidendrums may be propagated by removing the side shoots and growing them in singles or multiples as soon as roots have emerged from the sides of growths.

Epigenium (Gagnepain) 1932

Epigenium, like eria, belongs to the dendrobieae. The name is fairly modern and dissimilar to the originals, which were changed several times. The changes included sarcopodium (Lindley) and bulbophyllum (Reichenbach f.). The nature of the plants gave rise to this and it is recorded in various publications dating from early in the nineteenth century. The genus comprises some thirty or more species, most of which are uncommon in collections.

The species occur in Indo-Asia and across to the Philippine islands, Java, Sumatra and Borneo. At one time *Dendrobium muricatum* (Finet), from New Caledonia, was also classified as an epigenium.

Three species are native to northern India and close countries and *Epigenium amplum* (Lindley) Summerhayes indicates how history has changed it several times. Originally described as *Dendrobium amplum* (Lindley) in 1830, it was also known as *Sarcopodium amplum* (Lindley), *Bulbophyllum amplum* (Lindley) Reichenbach f., *Callista amplum* (Lindley) Kuntze and *Katharinea ampla* (Lindley) A. D. Hawkes.

Most of the genus are classed as intermediate to warm-growing and many consider the most beautiful species *Epigenium acuminatum* (Rolfe) Summerhayes, which is illustrated in *Growing Orchids, Book Three.* It is also known occasionally as *Sarcopodium lyonii* (Ames) Summerhayes, and comes from various localities on the Philippine island of Luzon, where it grows at about 1000 to 2000 metres (3200 to 6500 feet).

Morphological variation produces plants with spaces between pseudo-bulbs of some centimetres, and others in which they are set closer. Each should be grown in the manner dictated by this. Treefern slabs will be found suitable for some species; others may be grown in containers and they select themselves for the method. Bark mixes are suitable for those grown in containers.

The Indian species *E. amplum* carries only single flowers from the apex of the pseudo-bulb; they are green-brown coloured and with deeper blotches. *E. acuminatum* is perhaps the most colourful with its pink flowers which have a deep rose central area.

Treatment is as for dendrobiums, with water and fertiliser applied rather sparingly unless the plants are in full flush of growth. Flowering should be in spring to early summer, but may be irregular in cultivated plants, particularly when transferred to the southern hemisphere and non-tropical climates. Over-watering at any time of the year should be avoided: it is better to keep the pseudo-bulbs slightly shrivelled rather than aim to produce fat, well-fed looking plants suitable for intermediate to warm climates.

Eria (Lindley) 1825

An infrequently grown yet large genus common to much of Indo-Asia and reaching down to the Pacific Ocean, with a large number of species in Niugini and down into New Caledonia. There are also several Australian species. Most of them suit intermediate climates.

The genus is allied to dendrobiums and some plants could be mistaken for that genus in their habit and appearance. Others, such as *Eria coronaria* (Lindley) Reichenbach f., have no pseudo-bulbs as such and are similar to some terrestrial genera in appearance. If they were freely available it is possible that more growers would appreciate their beauty and unusual flowers. While some flower from what appears to be terminal inflorescence, most have the habit of flowering from upper nodes of the pseudo-bulbs and usually from all of these at the one season, the pseudo-bulb then producing a new shoot and slowly deteriorating.

Cultivation of the genus commenced in 1825 with the introduction of *Eria flava* (Lindley) from northern India or Burma; it was flowered by a grower at Liverpool Botanic Gardens. In the same year *Eria stellata* (Lindley), the correct name of which is *Eria javanica* (Swartz) Blume, was flowered by William Cattley, the man who had the singular honour of flowering the first of the cattleyas. The correct name for *Eria flava* is *Eria pubescens* (Hooker f.) Lindley ex Steud. In those last few sentences is a condensed history of the nomenclature of many of the orchids of the time. Several authorities were responsible for cataloguing all the introductions from different parts of the world, and communication was slow and at times non-existent. The inevitable result is that there may still be adjustments made from time to time to lists of species and even genera which were collected, grown and flowered in the early years of the 19th century.

In Bartle Grant's *The Orchids of Burma*, a brief quote appears in relation to the key to eria species which explains some of the confusion and its cause: 'Perhaps the most polymorphic genus of Orchids, and very difficult of division into definable groups. Though never confounded with Dendrobium, the only available distinctive character between them is the oblong pollen of the latter genus and more or less pyriform of "Eria".' The quote is probably from Lindley.

Generally erias should be treated as epiphytic orchids, and growers will mostly choose potting material based on bark. As the flowering periods vary, the plants should be brought to maturity if possible and flowered in late spring or early summer, with alternative summer growth and early autumn flowering, followed by a period of dormancy and less moisture. Some may not flower regularly or in their correct season.

Eulophia (Robert Brown) 1823

A genus which grows in almost every continent except Europe and which, like habenaria, another almost world-wide genus, is little cultivated. Eulophias grow as terrestrial orchids, in appearance somewhat similar to phaius, and with the spike, sometimes quite tall, emanating from the base of the pseudo-bulbs. These pseudo-bulbs are frequently sub-surface and a new growth may be some centimetres tall as the flower spikes bloom.

In cultivation the genus should be treated as terrestrial, growing with the pseudo-bulbs partly or wholly sub-surface, although some growers embed the sometimes leafless bulbs only half-way.

Southern Africa has a number of species and they are very attractive when in flower.

Joyce Stewart, in her book *Orchids of Africa*, indicates that there are some 200 species in the genus. This book is beautifully illustrated by Esme Hennessy. Both are botanists of note. More than forty species grow in South Africa and a larger number still in Zambia.

Terrestrials of this type are never attractive to cultivators of species orchids and have been only infrequently noted by the author when looking over the plants benched at shows or in flower in collections. Terrestrial orchids are more at risk than epiphytes and lithophytes because of their position in the ecology of countries where they grow. Cultivation of land is the principal cause of destruction.

If species fanciers are attracted to eulophias in single specimens or as flasks of seedlings they should understand the nature of their habitats. In some ways the plants relate to Australian genera as some may produce flower spikes from the rhizome-like pseudo-bulbs without first growing leaves, which follow later from the rhizome as it extends. Some plants described by Joyce Stewart appear similar in their morphology to *Dipodium punctatum* (Smith) Robert Brown, with the flower spike protruding from the grassland or sandy terrain, followed by the leaves. It is frequently imagined that *Dipodium punctatum* is leafless, but those who have noted the spikes emerging from the floor of the eucalypt forest which is generally their habitat cannot fail to have noted the rosette tip, which is impermanent.

Eulophias generally have a dormant period which follows the stages of flowering, leafing and maturity. When the leaves are cast, the plants should be rested.

Potting material should be slightly enriched cymbidium type mix and it should be capable of reconstitution after the drying off in dormancy.

Australia has several species and possibly there are some which have not been noted. All occur in northern Australia and are neither plentiful as plants nor widespread, with the exception of one. Most have peculiar pseudo-bulbs, perhaps best identified by comparison with a common ginger root. The flower spike emanates from this pseudo-bulb and a leaf stalk follows. Flowering time of nearly all species world-wide is spring and summer.

The distribution of the genus in the Australasian region is curious. Niugini has several species, but it is possible more exist than those appearing in Schlechter's *Orchidaceae of German New Guinea*, one of the more complete records of Niugini species. The genus extends into Australia with five species listed in Clements' checklist, which is classed as a preliminary survey. There may be more. New Caledonia has three, one of which is also listed as Australian.

But the critical point is that so far as listed orchids are concerned New Zealand has none. Considering the Gondwana bond, it would appear that the migration route followed by the genus has been southward, in contrast to the inferred migration routes of other Australian genera. Not all authorities would agree with the interpretation, illustrated in *Growing Orchids, Book Four*.

Turning back a few pages, *Sanders' Orchid Guide*, the last reprint of which was in 1927, lists the genus as: 'A widely distributed genus of epiphytal or semi-epiphytal and terrestrial orchids. But few of the species are cultivated.' Most of the species were from warm to hot climates and this appears to be a source of information for growing them. They are essentially tropical orchids and need warm climates rather than artificially warmed environments.

For those interested in the genus a large collection could be made if formed around the African species, but it is very difficult to obtain plants. Many are illustrated in *Wild Orchids of Southern Africa*, by Joyce Stewart, H. P. Linder, E. A. Schelpe and A. V. Hall.

In 1823, in addition to *Calanthe veratrifolia* (Willdenow) Robert Brown (syn. *C. triplicata* (Willemet) Ames), *Eulophia gracilis* was sent to Britain by G. Don. This is one of the first indications of the genus being brought into that country and possibly among the first species to flower in that country. Over a long period very few more were to appear in cultivation, although they were grown correctly in suitable terrestrial mixes. Perhaps strangest of all, the genus apparently has never been the subject of hybridisation.

Galeandra (Lindley) 1830

A genus of medium-size plants confined to parts of Central America, some Caribbean islands and the northern states of South America. Its habitats are usually situated in humid, warm areas, the plants generally growing as epiphytes, very occasionally lithophytes and occasionally terrestrial in extremely wet regions at times during the annual cycle.

Some galeandras are ansellia-like in appearance, with four or five leaves, narrow and of moderate substance and rather impermanent. Others have somewhat different appearance, but pseudo-bulbs other than the ansellia-like growths are not present.

The flower stems emerge from the apex of the pseudo-bulbs, again similarly to ansellias, the flowers varying in number from species to species, in some opening consecutively up the stem and in others all opening at once. The flowers of most are somewhat like phaius, particularly *Phaius tankervilleae* (Aiton) Blume, but with a spur protruding from the base of the labellum. Colour varies from species to species, particularly the labellum. The petals and sepals are rather narrow and insignificant; they are principally brownish to lemon coloured.

Brazil has about twenty species and all would be considered tropical to sub-tropical orchids, needing minimum temperatures of about 20 degrees (68°F.), and tolerating much higher temperatures provided the humidity is also high, about 70 to 80 per cent. Most of the species flower in summer to autumn and the genus as a whole would be unsuited to climates which are dull and cold for half the year. The plants prefer bright conditions: in tropical climates shade cloth of about 50 per cent would probably be sufficient.

The genus is uncommon in collections, with the predominant species cultivated being *Galeandra devoniana* (Schomburgk) or *G. baueri* (Lindley).

Veitch states that Schomburgk sent to Lindley the following description of his finding *Galeandra devoniana* in Brazil:

> During our peregrinations we have seen this plant nowhere else than on the banks of the Rio Negro, a tributary of the Amazon, where in the neighbourhood of Barcellos we found it growing in large clusters on the trees which lined the river, sometimes on the *Mauritia aculeata*, or even on the ground where the soil consisted of vegetable mould. It was so luxuriant in growth that some of the large clusters of stems which sprouted from a common root were from 10 to 12 feet [3 to 3.5 metres] in circumference. The stems were often 5–6 feet [1.5 metres] high; at the lower part almost of a purple appearance but changing into green higher up. As the flower is not only larger than the generality of its tribe, but handsome, I availed myself of this opportunity of naming it in honour of the Duke of Devonshire, one of the most successful cultivators of this, one of the most interesting tribes among monocotyledonous plants.

That quotation has been repeated in numerous publications. The success of the duke in cultivating the plant should be gauged by the experience of many who followed him in the pastime. Galeandras have proved difficult to house with other orchids because of their needs. They have been grown and flowered for short periods but usually prove too much of a trial to grow into large clumps as seen by Schomburgk. The pseudo-bulbs, too, become much shorter in cultivation, usually growing to only about 30 to 50 centimetres (12 to 18 inches). It has been found also in other regions.

The root system is the cause of most failures and it is difficult to combine modern ingredients into a suitable potting material. Whatever is used should be similar to that used for calanthes, with fast-growing, rich material to boost growth and flowering, one following the other quite quickly. It should also be possible to leave galeandras undisturbed for long periods so that a suitable clump may be formed. Frequent repotting should be avoided, but if it becomes necessary the job should be tackled as soon as the need is noticed, otherwise the plant will quickly decline and probably be lost.

While the plants are seldom noted in collections, they grow well in humid, warm climates. If possible growers should try to pollinate flowers and have the seed raised to flask and preferably replated level. No doubt raising seedlings to flowering plants would be an arduous and time-consuming task, but with the paucity of imported plants, it is possible this will be the only means available to grow and flower galeandras outside their native countries.

The somewhat fusiforme pseudo-bulbs of *Galeandra devoniana* are similar to those of some Australian dendrobiums of the tropical sector, even down to the colouring and the foliage, although that of the galeandra is somewhat longer and slightly broader.

Fertilisers, considering the nature of the growth, should preferably be dilute liquid nutrients rather than slow-release solids, and care in frequency and duration of fertilising is needed.

The morphology of the genus as a whole is variable and should not be based entirely on the description of *Galeandra devoniana*. Generally there is little difference in recommendations so far as climate and potting materials are concerned. But mixed species should be carefully considered for tropical cultivation.

Gongora (Ruiz and Pavon) 1794

A small rather infrequently grown and confusing epiphytic genus. Confusing because the species *G. quinquenervis* (Ruiz and Pavon) has fourteen synonyms, some of which are also used as the correct names. This species is native to Central America, Venezuela, Brazil, Peru, Trinidad and possibly Colombia, which represents the total spread of the genus.

Albinism or perhaps lutinism, common to most genera, is also known in some species. There is no general agreement by authorities on names, and this is a disturbing feature, particularly when one tries to relate photographs of specimens to printed descriptions. The flowers, like those of stanhopeas, are strangely formed.

One of the first species to be brought to cultivation was *Gongora atropurpureum* (Hooker), originating in Trinidad; it was flowered in the Liverpool Botanic Gardens in 1825. It is now considered to be conspecific with *G. quinquenervis*, the authors in another source quoted as Ruiz, Lopez and Pavon.

Colour variation in species from region to region adds to the difficulties. This becomes more apparent when using colour photography and comparing transparencies with reproductions in some of the older books using lithographs such as Curtis's *Botanical Magazine*.

Records of the Royal Horticultural Society contain many examples of well-flowered plants. One produced by Sir Trevor Lawrence, president of the society in the early 1900s, was exhibited at a meeting and carried about twenty pendulous spikes. Little thought is needed to imagine the transport difficulties without considering the culture involved.

The genus should be treated as open-root orchids best suited by being mounted on slabs or placed in hanging baskets. Plants may be pot grown, but the material would need to be related to the climate. Bark should preferably not be used in sub-tropical to tropical regions: a harder stony material should be substituted.

Root systems of many species are occasionally aerial in the manner of some other genera. *Dendrobium speciosum* (Smith) is a good example, with new root growth at times vertical to about 20 centimetres (8 inches) and green-tipped when young. Some catasetums grow similar roots and this is illustrated in *Growing Orchids, Book Three*.

Such root systems are encouraged by climates with high humidity but not necessarily high rainfall. As the vertical system dries off at the tips, the plants go into flowering and then dormancy; some leaves may be cast and the pseudo-bulbs become ribbed.

This is a common phenomenon with gongoras and direct watering should be avoided until new growth shows. In this stage root action should be first noted; as this is difficult to see, the newer pseudo-bulbs should be watched for growth.

This stage should be fairly fast and flowering follows, generally in summer to autumn, but at times variable, according to cultivation.

Shrivelling or ribbing of the pseudo-bulbs in the genus as a whole is common and should not be regarded as a signal for watering. The moisture should be in the atmosphere and direct watering or fertilising of plants should not be regular until new growth is on the way. Otherwise only occasionally dip the base of the container to moisten it, or spray the plant very lightly if it is slab mounted.

Slow-release pellets are not recommended. The author is opposed to this form of fertilising for all epiphytic orchids, because there is too much likelihood that some residue of fertiliser will be left when the plants become dormant, and will harm the roots.

Gongoras, like many other Central and South American orchids, are on prohibited export lists and only divisions of existing plants in collections or seedlings are available. Species orchid cultivators should make pollinations, preferably using two plants and using the pollen from both. It should also be tried on more than one flower on the raceme so as to promote the chance of viable seed.

Species do not all flower regularly or in the correct season denoted for their habitat; in cultivation they may transfer from summer to autumn or spring to summer, depending on the climate in which the plants are grown. If the system is suitable the plant should produce two shoots from a lead, thus giving an increase in what would be the normal way of plants in the habitat.

The habit is usually for seedlings to develop into mature plants on rather low-growing situations such as forks of trees, often described by early collectors within easy reach. The plants grow in fairly bright situations and need good light in artificial culture. Their appearance is nearly that of bifrenarias, but the leaves are less substantial.

In semi-tropical or tropical climates, if the plants are suited, shade cloth of about 60 to 75 per cent would be ample. If they are grown hard the leaves are light green, almost yellow, and in this state they should be watched so that they do not burn.

Grammangis (Reichenbach f.) 1860

Some genera are comparatively rare in cultivation, this one among them. Native to Madagascar, it has few accompanying species in the genus and the taxonomy has been confused from the beginning. Originally named a grammatophyllum by Lindley, the author of its present name created a new genus for the group.

The species in Madagascar has only one to accompany it in the Malayan–Javanese region of the world. This is not the only genus which has such wide areas between individual and group members of its species. One which was previously credited with such a distance from the main members was *Amesiella philippinensis*, which was until recent years known as *Angraecum philippinense* (Ames).

Grammangis ellisii (Lindley) Reichenbach f. was first brought to the notice of the orchid community in 1859, when the Rev. Ellis first brought a plant to England from Madagascar and flowered it in his collection. He described it thus:

> Among the plants which I brought from Madagascar was a large-bulbed plant exactly like *Anguloa clowesii*, only the bulbs were square instead of round. I found it growing on a branch of a tree about the size of a man's leg and stretched over a stream at about 25 feet [7.5 metres] above the water.

The pseudo-bulbs are about 20 centimetres (8 inches) tall and square in outline, the leaves more like those of broad cymbidium foliage than of anguloas. The plants, coming from a climate such as that of Madagascar and considering the fact that they usually occur along the banks of streams, need high humidity and warmth that is infrequently used in cultivating most of our orchid species. Moderate shade should be used, probably in the order of 60 to 75 per cent woven cloth; it will be found that the leaves are subject to burning more quickly than those of cymbidiums. Temperatures should not fall below about 18 degrees (65°F.).

Containers should be relatively large, filled with coarse material containing humus such as dry leaf which will not deteriorate to mud as it decays. Open-root cultivation on mounts such as treefern or similar material, requires careful attention to new root growth to ensure that the mount is suitable. Alternatively, the plant may be grown in a wire basket with a lining of shade cloth of coarse weave, with the same potting material as for other containers.

Once growth and flowering have finished, plants should be rested until new action commences. The illustration on page 142 shows a plant flowering in southern Australia.

Grammatophyllum (Blume) 1825

A genus restricted to the south-eastern Asian region extending from Malaysia to the Solomon Islands, and relatively small in numbers. The members are all recommended as tropical to sub-tropical species and when glasshouse grown need bright conditions and temperatures in the warm to hot range of 25 to 30 degrees (77–95°F.) as a minimum. Some variation below this range may be tolerable but could restrict flowering.

Uncommon in cultivation, probably the species most often seen is *G. scriptum* (Blume), which has a wide distribution, including the southernmost extent of the habitats. It was noted in the Solomon Islands on Bougainville when the author grew it in a collection established there. Plants grew almost on the coastline. A peculiarity it shares with other genera and noted in regard to gongoras is the habit of plants growing vertical root systems about the perimeter of the clump. Plants grown by the author there and established in a matter of weeks also produced this feature.

In nature plants almost invariably occurred in the forks of trees and in this position were at times with other genera in a tangled mass. The morphology resembles cymbidium growth, although the leaves are stouter and shorter; unless the plants were known they could be mistaken for that genus.

Grammatophyllum speciosum (Blume) is among the largest orchid plants and this species was imported into England about 1850. It proved difficult to flower on stems or pseudo-bulbs which were 'from nine to ten feet [2.7 to 3 metres] high, the flower stems six feet [1.8 metres] long and the expanded flowers nearly six inches [15 centimetres] across.' Those of *G. scriptum* are mostly about 5 to 6 centimetres (2 inches) across.

Grammatophyllum scriptum is grown in northern Australia, where the climate suits it and the temperature seldom falls below the limits quoted.

Containers for cultivation should be large and the potting mixture coarse but rich in humus. They are not open-rooted epiphytes but may be grown that way on suitable mounts and a profuse root system encouraged. Bright conditions are essential: plants growing naturally are in full sunlight for some of the day.

Fertilisers may be of the slow-release type and this is one of few genera which the author regards as suited by this type of feeding. In the growth stages the plants need plenty of water to encourage fast growth.

Grammatophyllum scriptum (Blume) may also be known under the name of *G. papuana* (J. J. Smith).

Grobya (Lindley) 1835

Apparently isolated to Brazil, this small genus is widespread through that enormous land mass, growng in the southern states of Rio de Janeiro, Guanabara, Sao Paulo, Parana, Minas Geraes and Rio Grande do Sul. In this sense the three species are all southern Brazilian and perhaps may also grow in countries to the west of Brazil, for which information is sketchy.

The plants grow naturally in a climate which is hard to duplicate in artificial conditions. It is temperate yet humid, with a fairly constant rainfall which is light and at times misty, and with good air movement. Artificial climates seldom resemble these environments and we also tend to conglomerate plants of various genera in an average set of conditions which is adjusted to suit the greatest number. It is seldom that the full flowering potential of grobyas is reached, as they may carry twenty to thirty flowers on a raceme, while artificially grown plants may carry fifteen or a few more. The lowest temperature should be about 12 to 15 degrees (54–60°F).

The genus was based on *Grobya amherstiae* (Lindley), which was sent from Brazil to Montreal and flowered in 1834 in the collection of the Countess Amherst, although the genus owed its name to Groby, the estate of Lord Grey. He was 'a munificent patron of horticulture and a most zealous cultivator of orchideous epiphytes', and lived in Derby, England.

In the past this orchid has not attracted much attention from growers, but a stimulated interest in species orchids of many genera has brought it to light once again and divisions are now eagerly sought.

The plants have unusual pseudo-bulbs, as they are nearly spherical with the leaves growing from the enclosing sheaths and from the apex of the bulbs. These are rather impermanent, possibly due to incorrect conditions. The pseudo-bulbs may also produce new shoots from almost the largest diameter of the pseudo-bulbs and not always from the lower parts. They grow rather closely together. Shade should be about 60 to 75 per cent.

Cultivation of the most common species, *G. amherstiae* (Lindley), is best in containers with rather coarse but sufficiently close material to suit roots which are medium diameter and proliferate in the fashion of cymbidium roots. The plants originated as epiphytes but we convert them to suit our culture. They should not be too frequently disturbed.

There appears to be a definite resting period or dormancy after flowering, which is usually during summer to early autumn. Fertilisers should preferably be quickly dissolving solids such as light dressings of blood-and-bone or hoof-and-horn or liquids; they should be applied when the plants are growing and until the flowers have matured.

Habenaria (Willdenow) 1805

It is difficult to imagine why orchids which grow in every corner of the earth are so little cultivated. The genus has been known almost since the time of plant cultivation in the years before they were termed orchids. It may be because the genus is almost exclusively terrestrial and poses problems if represented by only one or two plants in collections of epiphytes.

That they are terrestrial orchids should not of itself give growers trouble because there are simple solutions to the problem of almost total dormancy for a long period during the year. A solution is always possible provided each grower understands the limitations of the environment and makes allowances for them.

It may be difficult to name the beginning of cultivation of the genus, but those entries best understood occurred somewhere in the latter half of the eighteenth century. The plant *Habenaria ciliaris*, a North American member of the genus, was referred to in Andrews' *Botanists' Repository*, first issued in 1797, and illustrated most probably by a woodcut block.

Before that period and for some years after it the genus was known by several names, such as plantanthera. In all there were more than twenty-six different names given to the genus. *Habenaria susannae* (Hooker f.) was Lindley's *Plantanthera susannae*. As a species this plant is known from Nepal in the west through to China and Java and other islands through the eastern archipelagos. The flower is almost pure white and about 6 to 7 centimetres (2½ inches) across, with the spur, a prominent feature of the genus, almost as long as the very long petals and sepals.

Its habitat in general is of sheltered, warm, moist surroundings. Duplication of local conditions enjoyed by each species, unless it is widespread, would need to take note of these primary requirements.

Among the first of the genus brought to England was the Canadian species *Habenaria fimbriata* about 1777, but the authority naming it has not been recorded in most books containing such information.

As with most terrestrial orchids, the plant sheds leaves and flower stalk after flowering and the underground tuber fills out any multiplication and goes into a rest period until the weather and seasonal change brings it into life again. Flowering takes place in late spring, summer and frequently into autumn, depending on the species and its origin.

Post-flowering treatment for terrestrials usually consists of slowly dehydrating the potting medium until it is almost but not quite dry. This applies particularly to habenarias: if they are totally dehydrated the tuber shrinks and frequently fails to rehabilitate and commence growing. Some means must be found to keep the tuber, which is frequently quite large, in a solid and non-shrivelled state while making sure that it is not over-watered and possibly decayed. Most of the stages are similar to the conditions applied to such orchids as disas (see pages 60–2) and in some ways the treatment of the two is parallel in seasonal requirements.

In addition to perfect drainage, a suitable potting material must be devised which will supply a supportive growing medium. Frequent repotting should be avoided, and is best delayed until the beginning of new growth is expected.

The double-potting system has many advantages. Although clay pots are dear and at times hard to get, they offer help in maintenance of the correct conditions of growing as well as dormant periods.

The tuber should be planted almost at the bottom of the potting material in the inner clay pot, which should be crocked to about a quarter of its depth. These crocks may be either pot shards or pieces of foam plastic which is inert and not harmful to roots. The plant, in its pot, is then put inside another pot, which may be either plastic or clay, with a light pad of moss put into it to fill the gap between the bottoms of the two pots. With this system, watering during the growing season can be less frequent than if only one pot was used, and in the dormant period water is applied only by dipping the pots to the depth of the moss in the outer pot and allowing the moisture to seep into the base of the inner pot. By removing the inner pot at any time the amount of moisture present in it is discernible by the feel and look of the pot. Even if it is suspected that it is dry, rubbing the fingers over it will probably indicate that it is moist.

Potting material is a matter for individual choice, but that recommended for disas is suitable at least for a start.

It is not easy to obtain plants of habenarias, but since they grow in almost every country in the world, some species will be found to suit every climate.

Haraella (Kudo) 1930

This genus is among the minor genera of the Asiatic area, and has only two species. Its affiliation is with saccolabiums and probably the two could easily cross-pollinate. Natives of Taiwan, they are infrequently noticed in orchid collections and in most instances would be in those of people who concentrate on species orchids, regardless of the size of flowers.

There are many of these smaller orchids which could be nominated as worthwhile subjects for such collections as the smaller saccolabium-like group. The foliage of haraellas is similar to that of some phalaenopsis as well as saccolabiums, but is rather thinner in substance than those orchids and liable to burn easily in exposed conditions.

In character with its origins in Taiwan, the plant flowers several times a year in favourable conditions and this could possibly be expected if grown in the right artificial environment.

It could be thought that the plants are large, but in reality they are miniature, with leaves 5 centimetres (2 inches) long and about 2 centimetres (3/4 inch) wide. The flowers are about 9 to 12 millimetres (1/2 inch) across, lemon or lemon-green, with a dark red-brown patch on the labellum.

In cultivation these small orchids should be slab-mounted, preferably on hard wood or cork which is suitable for the roots. If they show a tendency to avoid the mount or do not attach to it the plant should be removed and tried on different wood. In general, well-weathered sections should be suitable, but the roots soon prove this.

Fertilisers for these tropical miniatures are usually individual and they should not be treated as ordinary genera grown in bulk. The smaller the plant the more likely it is to be forgotten or to remain dry when it should be moist. *H. odorata* (Kudo) grows in the mountains of northern and central Taiwan at about 300 to 900 metres (1000 to 3000 feet), mostly in light forest and semi-shade.

In cultivation the temperature should be maintained above 12 degrees (54°F.) through the coldest parts of the year. The growing season is generally in summer; provided the shade is correct at about 75 per cent, the plant should grow and flower almost simultaneously.

It would be a mistake to regard the orchid as anything other than a tropical type, but it does not favour humid still heat. Good air flow is a necessity.

Hexisia (Lindley) 1834

A genus with very confused recognition, some authorities limiting its occurrence to certain parts of Central America and the northern part of South America, while others regard it as almost pandemic. Apparently the species *H. bidentata* (Lindley) is known from Peru to Guyana, with occurrence in all the intervening areas where such orchids may grow. It is also widely known in the Central American countries, although not having credit in some.

There are few species in the genus. Over the period of its known history it has had a number of taxonomic changes, which appears usual in most genera. The species *H. bidentata* (Lindley) has known similar treatment.

The genus has a peculiarity, not generally known in orchid growth, of elongating the flowering stems into new growths rather than continually initiating new growths from the rhizome or the lower part of existing growths. It has this in common with other orchids such as flickingeria. The new growth flowers in its turn, and again carries on a similar process. There is a characteristic weakening of the growths as the stem elongates, but as the plant is not simply monopodial, it also produces new shoots from lower parts.

As a plant this form of growth in time becomes untidy, and it is more easily maintained if it is mounted on a slab of fibrous material which carries the root system. It should not be grown in a pot, no matter how small. If grown in a basket, it may be found after a few years that it needs handling to reduce some of the growth material and to get new 'breaks' in the stems.

The flowers, brilliantly crystalline red, are borne in groups of three or occasionally more in autumn, winter or spring. The plant grows again after flowering.

In good conditions it may flower more than once in the annual cycle, but temperature and good air flow must be carefully regulated. Recommmended minimum is about 12 to 15 degrees (54–60°F.) through the cooler portion of the year; there is almost no limit on the upper range provided humidity is sufficient.

As an unusual addition to a collection it can be recommended for semi-tropical and tropical climates, with good tolerance for bright light and warmth. If grown in an enclosure, about 50 to 60 per cent shade is recommended. Fertilisers are more easily applied as liquid sprays to the base of the plant and the potting or mounting material. So-called foliar feeding may induce fungus infections of flower base and growth initiation points.

Huntleya (Bateman, ex Lindley) 1837

A genus of few members with many confusing changes in nomenclature. At one time it was thought to be exclusively confined to part of the Central American and South American areas, but it has since been found in many isolated places, including the island of Trinidad. Unfortunately the material collected fell into different hands and this accounted for much confusion. It is not plentiful.

The best known of the genus is *Huntleya meleagris* (Lindley). Only once or twice has the author seen it in flower, first about 1960, and it is doubtful if many plants are at present in cultivation. According to reports at that time it was not an easy orchid to locate, unlike some other genera in the South American region. Although originally found at about 1000 metres (3250 feet), the plant should not be regarded as a simple project in cultivation.

It is reminiscent of angraecums in its form, although the foliage is not thick; it is rather insubstantial, about 50 centimetres (20 inches) long, rather broad and distichous. The flower stem emerges, vanda-like, from the axil of the leaf. The flowers are held on stems about the length of the leaves, usually two or at most three on each stem. The substance of the flowers is coarse and crisp and the colours variable.

The plant was described by a resident of Trinidad as often found growing, when they were fortunate enough to find it, with plants of *Brassia allenii* (L. O. Williams ex Schweinfurth). This last-named orchid has now been transferred to the *Ada aurantiaca* group and adds another orchid to what was once regarded as an almost monotypic genus. Plants of *Huntleya meleagris* were never plentiful, usually consisting of isolated specimens.

In cultivation it may be difficult to work out a program for it. An open slat basket or a wire basket lined with fibre will suit it. The root system when in health resembles that of vandas or angraecums and can be quickly lost if incorrect environments are created. It does not mix well with other genera, but it could be tried in cattleya conditions with a minimum temperature of about 12 to 15 degrees (54–60°F), and medium light intensity of about 60 to 75 per cent.

The leaf system should not be sprayed too frequently and never late in the day when water could lodge in the leaf axils and not disperse before the plant cooled in the evening.

Fertilisers should be liquid, with only quick-dissolving solids provided in the growing period. When root growth is active, a top-dressing of growing sphagnum moss could be helpful.

Laelia (Lindley) 1831

Closely allied to cattleyas and other genera of Central and South American orchids, laelias are a numerous genus and with marked dissimilarities. As a group they are among the most common of cultivated genera and could be termed temperate climate orchids. Even those considered cool-growing species will accommodate to temperate conditions. Most of these so-called cool-growing laelias, including the best known and possibly largest-flowered of the species, *Laelia anceps* (Lindley), will succeed in temperate climates.

There is a group which is considered warm-growing, and this includes what were at one time named schomburgkias. Most of those left within that genus possibly would grow in temperate climates, outdoor or in glasshouses, but could prove difficult to flower.

Laelias may be considered a Brazilian species because of the numbers growing in that country. Several are of recent origin so far as names are concerned, although these have until recent taxonomic changes been included with other species as varieties. There is another way of looking at that, however, and that is the penchant of taxonomists for changing things to impress their names on the catalogue.

In one of the latest of these catalogues of Brazilian laelias, that of Pabst and Dungs, forty-six are listed in several sections. Some species are listed as small and others as medium, but none is as large as *Laelia purpurata* (Lindley and Paxton), a unifoliate cattleya-like plant and one with tremendous influence on orchid growing. *Laelia purpurata* is confined in habitat to southern Brazil and almost endless variations have been catalogued. Most cultivators of species orchids at some time in their growing careers should have tried to grow *Laelia purpurata*. The operative word is 'try', because so many fail to flower this beautiful warm-growing orchid. It is one of those which could be more easily cultivated in those conditions than in temperate climates.

The subdivision of the genus into groups is interesting because the normal differences in flower colour, growth height, stem length and other singular features appear to have some part in determining the groups or sections, as they are termed.

At least two Brazilian species will grow in cool conditions — *Laelia pumila* (Hooker) Reichenbach f. and *Laelia sincorana* (Schlechter). Distinguishing characteristics between them are confused, and often one is bought as the other. The characteristics are in the opinion of the author that *Laelia sincorana* has taller pseudo-bulbs which are similar to those of the *Cattleya labiata* (Lindley) group and the flower is also smilar to those of cattleyas (see page 149.) *Laelia pumila* (Hooker) appears to have shorter, ovoid pseudo-bulbs and shorter, thicker foliage.

Laelia cinnabarina (Bateman ex Lindley) is another species which is outstanding and has contributed much to orchid culture. It is considered a native only of Brazil but its presence in at least one neighbouring state has half-credence. It is true to name, with brilliant orange-red flowers on tall stems and again on elongated pseudo-bulbs which are quite purple.

Laelia tenebrosa (Rolfe), has a widespread habitat perhaps extending into adjoining states. It is sometimes of dark hue and at others carries apricot-tinged golden blooms. It is a separate and very beautiful subject, rare and at times difficult to flower in artificial cultivation.

In recognition of the fact that new orchids are being discovered quite frequently, one of the beautiful small laelias, *L. milleri* (Blume) was brought to the notice of growers in comparatively recent times.

Another group of species said to be cool-growing orchids exists in the Mexican section. Principal member of the group is *Laelia anceps* (Lindley), which includes many varieties of typical shape and colour. This orchid will grow in quite cold conditions, tolerating occasional short periods of about 6 degrees (43°F.).

The second major species in what appears to be an allied section is *Laelia autumnalis* (Lindley). According to many authorities it is considered to have two major parts, the other *Laelia gouldiana* (Reichenbach f.). It is difficult to get an ultimate authority for inclusion of the two into an entity and this is best left to growers to decide for themselves. There is little doubt about the division on visible identity (see the illustrations in *Growing Orchids, Book Two*).

Other laelias are native to Mexico and Central American states. Perhaps the best of these is *Laelia speciosa* (Humboldt, Bonpland and Kunth). It is a difficult orchid to grow; it needs temperate environments and is most unusual in orchid collections. It has been grown by the author in cool conditions; it refused to flower for a number of years and then in a good season grew and flowered quite well.

As a genus laelias require open-root cultivation. This is not always possible: if they are pot grown the material should be open and free draining, not necessarily to contain a rather wandering root system, but rather to hold and support the plant. It will be found difficult to confine the Mexican laelias to pots for very long, and any movement means root disturbance. None like this form of cultivation.

If plants are grown on mounts they should preferably be of treefern or similar material. The roots will proliferate into this far better after the initial tying-on if the slab of fern is watered only on the back for the first few months. Once the roots have penetrated the fibre, water or fertiliser can be sprayed on the top of the mount rather than on the front, so that it soaks down to the base of the plant.

Light is vital to most orchids, but for laelias it is paramount. Shade should not be too dense once the growing period has commenced, but if it is too bright there is always a danger that the new shoots will burn. This usually means lost growth and flowers for a year. About 60 per cent shade is sufficient; with experience it can be lightened toward the flowering period in order to ripen growths and start the flower buds.

It is important in laelia culture to allow a definite rest period once flowering has ended. This dormancy could be for as long as six months, particularly with plants that flower in early winter, like the Mexican laelias. They may not commence growth until December, depending on the type of climate in which they live. Do not water the plants until new root tips are visible on the older roots or new tips emerge from the rhizome just below the newly matured pseudo-bulbs. Moisten them only occasionally.

Smaller species such as the Brazilian *Laelia lucasiana* (Rolfe) are better kept in pots, as they are neat, closely growing species with little space between the pseudo-bulbs. Potting material should be free-draining and yet easy to moisten after the dormant rest period. This will mostly match that of slab-mounted plants but may be more seasonal. Each plant should be treated individually, especially with regard to watering during the summer.

Fertilisers for laelias should be liquid, fabricated to suit epiphytic genera and applied only during the growing period and the lead-up to flowering. Once dormant the plants should be left unfertilised. Remember the golden rule for watering — too little is preferable to too much. Water thoroughly once a week in the relevant periods.

Like cattleyas, laelias are in numbers that warrant the formation of a species collection in their own right. The listed species are of sufficient interest and variety to make the genus the base of a collection; many other orchids could be included, particularly epiphytic orchids rather than species such as paphiopedilums.

The states of the main country of origin, Brazil, give some guidance to the nature of the environments in which they grow. It should be remembered that some species are lithophytic as well as epiphytic; plants which grow in stony environments are usually hardier than purely epiphytic species.

Some basic cultural problems arise from this. They consist principally of some modification of growing techniques in the dormant or resting period, which should be dry in daytime yet followed by a moist and humid atmosphere at night. The dryer part of the year, wherever the plants are grown, is not always the coolest part, so this must be counteracted in cultural patterns. These patterns may take a few years to work out, but they will be followed naturally by better flowering and prolonged root activity rather than the loss of roots and leaves as part of a subnormal annual program.

The following is a list of species suitable for glasshouse cultivation rather than outdoor growing in cool climates. Those which will tolerate this cool environment will also suit less harsh climates.

L. alaorii (Brieger and Bicalho), from the Bahia, Brazil, region.

L. dayana (Reichenbach f.), Minas Geraes.

L. *pumila* (Hooker), Reichenbach f., Bahia.
L. *sincorana* (Schlechter), Bahia.

Some of these plants may also be known by synonyms which are not included here. Outdoor cultivation is possible in temperate climates, but protection is preferable, and the temperature should not fall below 12 to 15 degrees (54–60°F.).

The following are recommended as temperate climate species:

L. *jongheana* (Reichenbach f.), Minas Geraes.
L. *praestans* (Reichenbach f.), Minas Geraes.
L. *cattleyodes* (Barbosa Rodrigues), Rio de Janeiro.
L. *lundii* (Reichenbach f.), Minas Geraes, Goias, District Federal.
L. *brevicaulis* (H. G. Jones), Espirito Santo.
L. *harpophylla* (Reichenbach f.), Minas Geraes, Espirito Santo.
L. *kautskyi* (Pabst), Espirito Santo.
L. *caulescens* (Lindley), Minas Geraes.
L. *crispata* (Thunberg) Garay, Minas Geraes.
L. *crispilabia* (A. Richard ex Warner), Minas Geraes.
L. *mantiqueirae* (Pabst), Minas Geraes.
L. *longipes* (Reichenbach f.), Minas Geraes.
L. *pfisteri* (Pabst and Senghas), Bahia.
L. *ghillanyi* (Pabst), Minas Geraes.
L. *liliputana* (Pabst), Minas Geraes.
L. *lucasiana* (Rolfe), Minas Geraes.
L. *milletii* (St. Leger), origin unknown.
L. *reginae* (Pabst), Minas Geraes.
L. *angereri* (Pabst), Minas Geraes.
L. *bahiensis* (Schlechter), Bahia.
L. *blumenscheinii* (Pabst), Espirito Santo.
L. *briegeri* (Blum.), Minas Geraes.
L. *cinnabarina* (Bateman), Rio de Janeiro, Minas Geraes, Sao Paulo.
L. *cinnamomea* (Reichenbach f.), Minas Geraes.
L. *endsfeldzii* (Pabst), Minas Geraes.
L. *flava* (Lindley), Minas Geraes.
L. *gloedeniana* (Hoehne), Espirito Santo.
L. *macrobulbon* (Pabst), Espirito Santo.
L. *milleri* (Blum.), Minas Geraes.
L. *mixta* (Hoehne), Espirito Santo.
L. *bradei* (Pabst), Minas Geraes.
L. *esalqueana* (Blum.), Minas Geraes.
L. *itambana* (Pabst), Minas Geraes.

The regions quoted here are west and north of Rio de Janeiro except for Bahia, which is in the northern part of Brazil. In general the regions could be said to lie between the parallels of 15 and 25 degrees south.

The rest of South America and Central America is relatively poor in the laelia genus compared with Brazil and those dealt with in the text are the major part of the remainder.

A general idea of the climates in which the species grow was given in *Growing Orchids, Book Two*, which deals with cattleyas and other epiphytes. Most of the Brazilian laelias fall into the intermediate group, with an indefinite line dividing the species into cultural tables.

Older sources were used in compiling the list of species rather than the 'updated' versions which have more place in recognition of synonymy than replacing accepted nomenclature for the average grower.

Lemboglossum (Halbinger) 1984

A new genus, reflecting the difficulties facing the modern grower. While there can be little dispute with the motives which prompt botanists to clear out the shelves now and then and transfer plants into different categories to those they have occupied for many years, it may take a couple of generations for growers to alter the labels attached to their plants.

This genus is typical of the problem. Originally the species were referred to the odontoglossum complex, but considering the differences between South American species and Central American and Mexican species the author Halbinger thought the distinction should be reflected in the nomenclature.

Lemboglossum rossii (Lindley) Halbinger has been known and grown since the early years of the nineteenth century and is still included in most collections as *Odontoglossum rossii* (Lindley). It defeats many growers who are unable to duplicate the environment in which it thrives. Unlike the South American odontoglossums, it has certain periods of dryness, not approaching dehydration, in its annual cycle.

This delicate balance may be achieved by suspending direct watering for periods up to several weeks, yet retaining the moisture at the root system by double potting (see page 81). In a species orchid collection or one devoted largely to species, the strange mixture of genera must have some method to make sure that plants which go through the stages of *Lemboglossum rossii* are conditioned in an environment which will take the worry out of collective growing.

This orchid, once the pseudo-bulbs have matured and flowered, is prone to lose the entire root system if incorrect growing procedures allow the roots to remain wet. Moistness is totally different from free water which may remain in pots for days. It is always possible to know the condition of the inside pot by lifting it out. Moisture apparent on the outer surface indicates that the plant is in good health. If dry, the pot should be returned to the inner position it occupies and the base stood in 4 centimetres (1½ inches) of water for ten minutes. The alternative is to soak the moss pad and return the other pot without watering it. The moisture in the moss will flow upward to keep the inner pot moist.

Do not be in too much hurry to water these plants directly until new growth is well on the way and the plant is developing new roots.

Other Mexican orchids brought into this small genus are *Lemboglossum cervantesii* (La Llave and Lexarza) Halbinger, which may prove more trying in cultivation than *L. rossii* and for the same causes. It is somewhat similar morphologically to *L. rossii*, and is listed among the flora of Guatemala. Similarly treated is *L. cordatum* (Lindley) Halbinger.

The species *L. bictoniense* is rather different from the others, having a quite tall spike of flowers with up to fifteen or twenty blooms and some would say more attractive colouring. *Lemboglossum bictoniense* owed its name to James Bateman, one of the greatest supporters for species orchid culture as distinct from hybrids, who once derided the efforts of hybridists as a waste of time and beautiful flowers.

In all, six or more species are included in the genus, with *L. apterum* (La Llave and Lexarza) Halbinger and *L. maculatum* (La Llave and Lexarza) Halbinger included. These species are all native to Western Mexico.

Lemboglossum rossii is readily distinguishable by its intense marking, the contrasting white or pink labellum throwing the flower into relief. *L. cervantesii* is usually more delicately marked, with the spots of red-brown inclined to be centralised on petals and sepals. More vividly coloured varieties have been known, with the spotting almost covering the whole of the flower, tapering off toward the ends of petals and sepals into fine red markings.

Other orchids may be moved into the genus in the future, with others from adjacent states being added. The odontoglossum complex was at one period very broadly based; over the last fifty years orchids have been removed from it and new generic names coined to define them.

Inclusion of some genera in lists of orchid flora of Central American states is frequently disputed and may be based on reports of finding isolated plants which were later over-collected or obliterated from the country by other means such as timber-getting and clearing. It is impossible to estimate the number of species lost in this way each year, but it is considerable.

Lemboglossums should be grown in the mix recommended for them over the years, usually compounded of bark and drainage material. I have always used and advocated inclusion of oak leaf in its dried, pre-decayed state in small proportion and this is designed to give the plants the material in which they thrive as epiphytes.

I use the double potting system advocated above and have found it easy to observe, and efficient in promoting both growth and flowering.

Leptotes (Lindley) 1835

A Brazilian genus which possibly occurs also in Paraguay and Bolivia. Originally found in the same area as *Cattleya loddigesii* (Lindley) and almost certainly brought or sent to England with that orchid. *L. bicolor* (Lindley) was first flowered by Mrs Harrison, notable for her culture and flowering of new species from the Rio de Janeiro region.

Originally about seven species were recorded, but from the date of introduction these have been whittled down to the present number of three, possibly four.

Leptotes bicolor (Lindley) var. *glaucophylla* came to England from Gardner, the man who supposedly found *Cattleya labiata* (Lindley) growing in the Organ Mountains in much the same area. It is not as widely known and grown as its beauty and ease of culture in intermediate surroundings merits. It is a widespread species, recorded as occurring in Bahia, in northern Brazil, and through intervening territories to southern Brazil.

Leptotes unicolor, smaller than *L. bicolor* and pale coloured, grows in Minas Geraes and Parana and is said also to occur in Argentina. It is seldom noted in collections and appears only at rare intervals as a cultivated orchid.

Leptotes tenuis (Reichenbach f.) is also a miniature orchid bearing single pink flowers on the apex of the pseudo-bulbs, and also not freely cultivated. It has the synonym *L. minuta* (Rolfe). Plants of this species grow only about 5 or 6 centimetres high (3 inches). It was at one time supposed to be a variety of *L. bicolor*, flowered originally by John Day, whose collection was principally of species bought unflowered and beautifully portrayed by him in colour. The collector responsible for sending it to Messrs Low and Co. was Blunt, who was also responsible for other finds.

According to some authorities on culture, the species should grow well in intermediate climates, but others consider they are suitable for cool environments. They should not be subjected to cold conditions below about 12 to 15 degrees (about 54 to 60 degrees Fahrenheit).

The plants may be grown either as slab-mounted specimens or in pots. Slab-mounting makes it easy to observe the root system for growth or dormancy and with these relatively unknown species, this is an advantage.

For plants selected for potting, the material should be a judicious mix of fine and coarse bark up to about 5 millimetres (¼ inch) in size, to which some coarse gravel has been added to prevent clogging. Watering should be only sufficient to prevent shrivelling, and fertilisers should be applied only when the plants are growing and flowering. Shade should not be dense, but about the same as for cattleyas.

Ludisia (A. Richard) 1825

A genus of few members which has been in cultivation for about 150 years. Its history is confused because the species *L. discolor* (Lindley) has been known under various synonyms and has been considered at times as separable into entities. Like other orchids with beautiful foliage, it is grown for this feature more than for its flowers. These in comparison with the leaves are rather nondescript and mono-coloured.

Veitch used *L. discolor*, then known as *Haemaria discolor*, with the related genus and species *Macodes petiola* (Blume) Lindley in 1861 to produce the hybrid *Macomaria* Veitchii. This occurred in an age when experimental cross-pollination was in its infancy, and it is doubtful if the hybrid has survived.

Australia has a closely related species in *Anoectochilus yatesae* (Bailey), the genus named by Robert Brown in 1813. This orchid has its habitats in tropical Queensland, always near the banks of streams and rivers or other moist ground, usually in well-shaded positions receiving almost no direct sunlight.

The foliage of *Ludisia discolor* is usually deep purple-red almost to black, with reticulations of clear red glittering lines running through in the positions of the normal veins of leaves. These leaves usually grow up to six in number on well-grown plants and the flowers are produced from spikes emanating from the tops of the growths. They are white, with a bright yellow tip on the column, about 2 centimetres (¾ inch) across and opening consecutively up the stem, which elongates as the flower head develops.

Plants propagate from offsets from the main stem, mostly on the lower sections or from the rooting base of the stem, which has no pseudo-bulb system to maintain it. As the growths mature and flower the leaves remain in place; as they age they may turn bright red and then orange, giving the plant a fiery appearance which is most attractive. In cultivation they should never be allowed to dry out, as this has the effect of robbing the plant of vitality and affecting its growth pattern. Flowering period is usually spring to early summer, but may extend into late summer occasionally, according to the culture pattern.

Potting mixtures should be able to supply the plant's needs in the one pot for some years. Part could be fibrous, part of decaying material such as oak or other durable leaves and part moisture-holding material such as peat-moss. If this is used the drainage should be good but the material holding continuous moisture. Shade the plants well, with up to 75 per cent shade cloth or protection of other types.

Lycaste (Lindley) 1843

An unusual genus because of its widespread habitats and distinctive species. The genus is to be found in almost every state of Central America, Colombia, Peru, Venezuela, Brazil and other South American locations.

In cultivation the principal species is *Lycaste skinneri* (Bateman ex Lindley), syn. *Lycaste virginalis* (Scheidweiler). Like species in other genera, it is diverse in colour and form, varying from pure white through to deepest rose-pinks. It has always been a most popular orchid, but defeats the efforts of some growers.

The genus has more than twenty members, most of which are cultivated in collections in all countries. The principal requirement is to sort the species into those which will tolerate cool conditions, of which there are a few, and those which need a little more attention to temperature and refuse to grow well and flower in minimum temperatures below 15 degrees (60°F.). The upper range is also important, and the genus generally will not grow or flower in sub-tropical or tropical environments.

Those which prefer cool conditions are based on *Lycaste deppei* (Loddiges) Lindley, which will tolerate temperatures as low as 8 degrees (46°F.) in the cooler months of the year because the plants are then dormant. *L. deppei* is one of the easiest of orchids to grow and should be in all beginners' collections in cool climates.

Generally the yellow-green and green species need warmer temperatures than other colours, but it is not a division which should be relied on when making selections. The term warm should not be misinterpreted to mean sub-tropical or tropical.

Potting material usually provided for cymbidiums should he coarsened up a little with graded pine bark to suit most of the genus. It should be provided with adequate drainage which allows water to flow through the pot and leave the material only moist.

Watering is a seasonal matter and should be approached according to the state of growth and flowering of plants. When in active growth, most species can be allowed to become 'ribby' on older pseudo-bulbs, which will quickly pick up condition as the plant flowers and begins to mature new growths. Fertilisers should preferably be liquid, and shading not dense but sufficient to prevent burning. Do not water the plants over the leaves while in growth and preferably not while in dormancy. Plants should not defoliate, but some may do so when coming into flower.

Masdevallia (Ruiz and Pavon) 1794

A genus which has been in and out of fashion over the years since its first inclusion as a cultivated plant. In the early years of the twentieth century about 185 species had been catalogued and that number was slowly added to. The genus became a popular cultivated orchid again in the 1980s and several fanciers in Britain and Europe, together with a few in North America, undertook hybridising on a minor scale.

In the late years of the nineteenth century the genus was grown by a small number of fanciers who produced a few hybrids, some of which are still popular more than one hundred years later.

The genus is taken to be cool-growing, but one cannot put all the species into that category: intermediate climates appear to suit many species. There are exceptions, but these, despite origins at elevations of 2000 to 3000 metres (6500 to 9800 feet), do not thrive when grown in temperatures which for some time fall below 10 degrees (50°F.).

The history of the genus can be gauged from the date attached to it by Ruiz and Pavon, Spanish botanists and authors of the work *Flora Peruviana*. It is native to northern areas of South America in the states of Peru, Ecuador, Colombia (which is possibly the richest area), Brazil and Venezuela. The largest flowers are not necessarily the most beautiful, an honour which must be awarded to the species *Masdevallia coccinea* (Lindley), *M. veitchiana* (Reichenbach f.) and a choice between *M. ignea* (Reichenbach f.) and one or two others. *Masdevallia ignea* is possibly more frequently known by the name *M. militaris* (Reichenbach f.).

The extension of cultivation opened the way for exploration for new species and this is still being carried out towards the end of the twentieth century. Numbers have been added to the list and the total appears to be about 350 to 400 species. Many of these could be classed as of botanical interest only. But for growers interested in adding to collections or concentrating on species cultivation, the relative size of the flowers is unimportant.

The genus masdevallia has recently been subdivided into sections and new genera have been created. Botanists and taxonomists found inclusion of some forms of the flowers under a general heading of masdevallia inapt. The genera trisetella, dracula and dryadella were created, principally by Dr Luer, to classify this group which is so large, and diverse in its morphology.

The genus extends into Central America, although not in the numbers found in South America, tapering off into Mexico with only a few members. In Panama are found representatives of the new genera, with the listing of *Dryadella simula* (Reichenbach f.) Luer (1978). Also in the Panama listing appears *Triaristella dressleri* (Luer) (1978), one of four in the newly created genus trisetella.

Another species is *Dracula erythrochaete* (Reichenbach f.) Luer (1978), one of a number removed from masdevallia into that new genus, following appreciation of characteristics which are inconsistent with listing under the older name.

In cultivation there also appears to be need for different treatment for orchids in these new genera. The foliage of *Dracula erythrochaete*, for example, is considerably thinner in substance than that of masdevallias and needs more shade; also, it appears, a cooler temperature and environment. A common fault in cultivating this section is that the foliage dies from the tips downward until the leaf has little left of its original length.

Masdevallias as a genus usually have substantial foliage, although it is easily damaged by dryness, heat or sunburn through too little shade. Two factors are vitally important: shade of no less than 75 per cent and constant moistness about the root system. Temperature is a matter for concern: few if any of the genus are suitable for intermediate culture in its upper limits. The preferable maximum temperature at root level is about 15 to 20 degrees (60–68°F.). This equates with that of a cool to moderate daytime environment. The night temperature may be considerably lower, but should not fall below about 7 or 8 degrees (45°F.).

These temperatures have no relevance to habitat temperatures for most masdevallias, which grow in lower temperatures during the night and in very humid conditions. The morning brings wind and sun, which quickly dries the foliage and warms the environment even at the high altitudes at which the plants grow. There are naturally variations from these figures, both in nature and in cultivation, but in an effort to work out a set of conditions an average has to be struck to accommodate a mixed group of orchids other than masdevallias.

While it is scarcely possible to maintain these temperatures and conditions in a climate which can quickly lose or gain warmth such as a sub-tropical region, it is wise to install equipment to counteract daytime fluctuations above the temperatures quoted.

Pots and potting are of vital consequence. The system described with reference to lemboglossums or odontoglossums will suit masdevallias or any of the subdivisions of that genus. It is preferable to use clay pots. In general pots for the genera should be small rather than large, even if the root system is squeezed. The plants should be understood for what they are: leaves and possibly a minor rhizome with no pseudo-bulbs to retain moisture for the plants to subsist on if allowed to become dry. Do not allow them to dry out at any time, although neither growth nor flowering may be taking place.

Watering should preferably be by means of the double-potting system (see page 81). The pad of moss in the bottom of the outer pot should never be allowed to dry out.

The type of benching on which masdevallias are grown will affect plant health. An open mesh bench, like that used for epiphytic orchids, is not as good as a solid surface under the pots, because the circulating air quickly dries out the moss pad.

For busy people it is comforting to know that double-potting takes the guesswork out of cultivation; there is little chance that the plants will dehydrate. It is possible to check whether the moss pad is wet at any time during day or night. The plant is in no danger of chilling with unwanted water circulating from top to bottom through the root system.

The root system is unusual because unlike most orchids it does not branch or proliferate. Once destroyed, leaf loss follows.

There is considerable variation in the lowest temperature limits for certain species. Some masdevallias are recommended for warm growing and this is because amongst the 300-odd species some originate in warm lowlands. Others come from considerable elevation and will not tolerate warmth beyond a certain level.

If possible understand the plants you buy, their origin and particularly the country they came from, even if information is sometimes hard to get.

Fertilisers, considering the nature of the plants, should be liquid and applied in weak form and frequently, as the plants are apparently always active.

Maxillaria (Ruiz and Pavon) 1794

A genus of plants which grow in Central America, South American states and some of the Caribbean islands. It is numerous, diverse, and polymorphic. Some species classified as maxillarias could be taken into new-genus status, but this would involve 'splitting', something I deprecate. Most maxillaria species thrive in fairly bright light.

The diversity in plants and flowers is exemplified in the illustrations of *Maxillaria sanderiana* (Reichenbach f.) on page 159 and *Maxillaria seidelii* (Pabst) on page 160. Like other species, these prefer intermediate warmth.

In only three instances has the genus been used for production of hybrids — once within the genus and twice as inter-generic pollinations. The former was registered in1930, with one parent *M. sanderiana*, the largest species and probably the most attractive, but difficult to maintain in many collections.

M. sanderiana is native to Ecuador, grows at a fairly high elevation and is supposed to be cool-growing according to some authorities. It appears better suited to minimum temperatures of 15 to 18 degrees (60–65°F).

As a genus, maxillarias are widespread in Brazil. More than eighty species are native to that country and more than seventy in Venezuela. Peru, a much smaller country than Brazil, has more than one hundred. *Maxillaria sanderiana* is also listed as Peruvian in some sources but not in others. There is little information of reliable quality.

Most confusion among growers is found with *Maxillaria picta* (Hooker) and *maxillaria porphyrostele* (Reichenbach f.). Both are Brazilian and easily confused. The basic difference is that on *M. picta* the red-purple blotching is on the outer surfaces of the petals and sepals, with bright canary yellow on the inside, whereas the outer surfaces are light yellow and the spotting and blotching on the inside of the petals of *Maxillaria porphyrostele*. There is a little variation in the two, and once recognised they are easily separated.

The genus extends right through the Central American states and into Mexico. In all more than three hundred species are listed. In general few of these are cultivated, but the genus has a history of cultivation going back to the early years of the nineteenth century.

Veitch records that until 1843 the group included lycastes, paphinia, colax, warria, scuticaria and others under maxillaria. Lindley revised the list and named these separate entities in the *Botanical Register* of 1843. In its original hotch-potch maxillaria resembled the genus epidendrum of Linnaeus, under which several now separate genera were grouped.

In a genus with more than three hundred members it is difficult to point out one more important than others, but several are worth mentioning.

Maxillaria variabilis, (Bateman ex Lindley), native to Mexico and several contiguous states and islands of the Caribbean, is different from the usual type like *M. picta* (Hooker), which have tightly packed pseudo-bulbs. *M. variabilis* has a scandent or climbing form of rhizome, similar to other species in the genus. From this form it can be inferred that the orchids with such habits grow on vertical or near-vertical surfaces and may be grown on slabs of cork, treefern or other mounts and hung vertically, preferably in positions which are well lit. This is one of the necessities for cultivating most of the genus correctly, using shading as low as 50 per cent and preferably where the plants get sunlight as early as possible in the day.

The other extreme is preferred by the largest-flowered species, *M. sanderiana* (Reichenbach f.). It is also difficult to grow, and compared with most of the genus its life in cultivation is nearly always short. The best way to grow the plant is in a slat or wire basket lined with fibrous material such as coconut fibre or the outer sheathing leaves of stag or elk horn ferns, with the inside root section of these ferns in blocks as the growing medium. Watering and ventilation are where most growers go wrong: the plant must be encouraged to grow roots and then to hold them through dormancy in conditions which are dry yet not to the point of aridity. Getting it to produce its growth and flowers in the correct period of the year is not easy. It should flower in late summer to autumn.

Maxillarias were brought to the notice of growers early in the nineteenth century, those attempting their culture even then finding the genus difficult. Most plants were grown in what were termed stoves, placed on the brickwork of these stoves rather than plunged into the beds of heated material. Probably the method was drawn from experience, which found that they were really hardy yet needing warmth somewhere near 12 to 15 degrees (54–60°F.) as a minimum.

One species, *M. nigrescens* (Lindley), will withstand temperatures as low as 6 to 8 degrees (42–48°F.) and yet flower in mid-winter, with the usual flowering time autumn to winter. Native to Venezuela, it grows at about 2000 metres (6500 feet) and upward and is illustrated on page 159.

Dormancy when not growing or flowering is essential, regardless of shrinkage of pseudo-bulbs.

With that in view, two systems of growing are possible: as potted plants or slab-mounted specimens. Whatever form is used, there should be provision for dehydrating the material for some part of the annual cycle and reconstituting a lively growing medium as required. The type of pot should conform to those used throughout the collection.

Potting material should be medium-coarse. A suitable mix would be one-half bark fragments of 1 centimetre (3/8 inch), half that size for a remaining quarter, and the remaining quarter scoria or rock of the same size. Bark, of course, dries out completely and must be capable of being re-wetted. Some growers put crocks in the lower portion of pots and this has much to recommend it. In the re-wetting process plants may be stood in shallow water reaching to the depth of crocking, allowing moisture to rise into the material from the bottom.

They should be left there for no more than half an hour at a time. After a few days of this treatment watering is resumed from the top when needed.

Airiness of the upper layers of potting material, particularly if the air is rising and moist, is the ideal system for reconstituting a root system that has not been used by plants during dormancy.

Slab-mounted plants should be watered from the top downward about every three days and there is little worry about surplus water. If the plants are mounted on treefern, the back of the slab should be watered thoroughly about once a week, allowing the moisture to soak in. The roots soon find it and new roots will search for it.

It is always advisable to keep plants to sizes which may be handled easily, unless production of a specimen plant is the aim. These plants, of course, are much easier to generate on slabs, preferably treefern, which can be superimposed over another thinner but larger slab as the plant progresses. Pot-grown specimens need careful handling because of root damage in transferring to larger pots and the risk of a tight root ball forming in the centre of the clump.

Light tolerance of nearly all the species is high and shade cloth of 50 to 75 per cent is adequate; choose the the minimum shade the plants will tolerate.

In general the environment should not be too humid, as the genus is principally dry-growing. This term 'dry-growing' should not be misunderstood as applying to the plants. Relative humidity should be about 60 per cent.

↑ *ACAMPE PAPILLOSA* (Lindley).
Native to Northern India and Burma.
Summer to autumn flowering.
Intermediate to moderately warm climate.

← *ACINETA CHRYSANTHA*
(Morren), syn. *Acineta densa* (Lindley)
Growing in outdoor conditions, mildly sub-tropical climate. Native to Costa Rica, Panama, about 2000 metres (6500 feet).
Summer flowering.

↑ *ADA ALLENII* (L. O. Williams ex Schweinfurth). Native to Panama, about 1000 metres (3250 feet), cloud forest. Sub-tropical, provided shade and humidity accurately gauged. Perhaps better suited in temperate climate, and probably incompatible with *Ada aurantiaca*.

↓ *AERANGIS CITRATA* (Thouars) Schlechter. Madagascan member of the angraecum complex. Prefers dry-type mounting or slat wooden basket. Intermediate climate, summer flowering.

← *AERANGIS SOMALENSIS* (Schlechter). Considering the size of the flower, this spur, common to all the angraecum complex, is probably the longest. It is a dry-growing species, and care must be taken to keep the flowers dry. Intermediate climate.

← *AERANTHES ARACHNITES* (Thouars) Lindley. Grows best in container. Flower spikes up to 2 metres (6½ feet) long, thin and wiry, emerging from base of plant. Madagascan species, prefers warm climate. Flowers late summer to autumn. Spikes coiled for photograph.

→ *AERANTHES ARACHNITES* (Thouars) Lindley. Single flower, about 5 centimetres (2 inches) across. Flowers periodic, usually in ones or twos.

↓ *AERIDES CRISPA* (Lindley). Common to India and Sri Lanka, usually about 1000 metres (3250 feet), flowering spring to summer. Vandaceous habit, prefers open-root system. Moderately humid, warm climate.

← *AMESIELLA PHILIPPINENSE* (Ames) Garay. Once regarded as Philippine extension of the angraecums. Light forest orchid, usually found between 1000 and 1500 metres (3250 to 5000 feet). Intermediate climate in artificial cultivation, flowers summer to autumn.

↓ *ANCISTROCHILUS THOMPSONIANUS* (Reichenbach f.) Rolfe. The name may be disputed, but was borne by the plant when imported. The flowers are about 6 centimetres (about 2½ inches) across. Grown in light subtropical climate. The probable synonym for this species is *Ancistrochilus rothschildianus* (O'Brien).

↑ *ANGULOA CLIFTONII* (Rolfe). This Colombian species is typical of the so-called 'Tulip Orchids'. The flowers usually have a delicate perfume and last for about ten days to two weeks in the correct environment. Most orchid flowers, even tropical species, last longer in cooler conditions.

→ *ANGULOA RUCKERI* (Lindley). Natural hybrids in the genus are known, usually distinctly traceable. This species is Colombian. Intermediate climate.

↑ *ANSELLIA AFRICANA* (Lindley). After many years of separation, some into species status, the various forms of this African orchid appear to have been consolidated into a single entity. It appears in habit a rather hard-looking orchid, is easy to grow in cool to intermediate temperatures and flowers in spring to summer.

↓ *ARACHNIS CLARKEI* (Reichenbach f.) J. J. Smith), syn. *Esmeralda clarkei* (Rolfe). Old names are hard to eradicate. This Himalayan species grows at altitudes of 2000 to 2500 metres (6500 to 8200 feet) yet is a warm-growing orchid, flowering late summer to autumn.

→ *ARPOPHYLLUM SPICATUM* (La Llave and Lexarza). A Central and South American species. Intermediate growing and with a liking for bright light when the growths are developing. The flower stems should be kept dry when beginning at the apex of growths.

↑ *ASCOCENTRUM CURVIFOLIUM* (Lindley) Schlechter. A native of a sub-tropical area of Burma and Thailand. A warm-growing species, free flowering, branching rather than continually elongating a single stem.

↑ *ARUNDINA GRAMINIFOLIA* (D. Don) Hochr., syn. *A. bambusifolia* (Lindley). A bamboo-like series of pseudo-bulbs or growths develops into a large clump in warm conditions. Flowering is seasonal, but it frequently flowers for much of the year.

↑ *ASCOCENTRUM MINIATUM* (Lindley) Schlechter. A large plant which has developed from a single monopodial growth. The flowering is prolific on such plants, but usually lack of accommodation in warmed enclosures is such that divisions are taken from such developments. Intermediate climate.

← *ASPASIA LUNATA* (Lindley). A widespread Brazilian species which grows in a warm climate; epiphytic and very occasionally lithophytic in aggregations of humus. Flowering summer to autumn.

↑ *BAPTISTONIA ECHINATA* (Barbosa Rodrigues). The author of this genus also that of the single species. It is allied to oncidiums. The flowering habit is pendulous, usually summer to autumn, and cultivation temperate to warm. The plant is oncidium-like, the flowers about 2 to 3 centimetres (¾ to 1½ inches) across, the labellum glossy and brilliant. A Brazilian orchid from the states of Rio de Janeiro and Sao Paulo.

↓ *BAPTISTONIA ECHINATA* (Barbosa Rodrigues), syn. *Epidendrum tetrapetalum* (Vell. non Jacq.).

← *BARKERIA SKINNERI* (Bateman ex Lindley), syn. *Epidendrum skinneri* (Bateman ex Lindley). A Central American species with a confused taxonomy. It is warm growing, and flowers late summer to autumn.

↓ *BIFRENARIA HARRISONIAE* (Hooker) Reichenbach f. The albino or clear colour form of this beautiful orchid. The form usually grown (cream coloured with a purple labellum) was illustrated in *Growing Orchids, Book Three*. The genus is South American, principally Brazilian, preferring moderately cool to intermediate climates.

↑ *BIFRENARIA INODORA* (Lindley). The short stems and clustered flowers are common to most of the genus. Brazilian.

↓ *BIFRENARIA WITTIGII* (Reichenbach f.) Hoehne. A species from a coastal district of Espirito Santo, north of Rio de Janeiro, probably needing intermediate conditions to grow and flower well.

↑ *BRASSAVOLA NODOSA* (Linnaeus) Lindley. A very hardy orchid but needing intermediate to warm conditions to grow and flower well. The terete foliage indicates something of the habitat: it grows in bright light as an epiphytic or lithophytic orchid, common to wide areas of Central and South America. Summer flowering.

↓ *BRASSIA KEILIANA* (Reichenbach f. ex Lindley). The colour contrast in the labellum, petals and sepals of this Venezuelan species mark it as one worth growing. An intermediate climate type coming from about 1200 to 1500 metres (4000 to 5000 feet) in moist forest.

↑ *BRASSIA VERRUCOSA* (Lindley). A Central American species, and perhaps mistakenly credited as South American. It is one of the most beautiful of the genus with its creamy white flowers marked in green. Summer to autumn flowering, intermediate climate.

↓ *BROUGHTONIA SANGUINEA* var. *ALBA* (Swartz) Robert Brown. This beautiful small Jamaican orchid needs a warm climate and is preferably grown open-rooted. Usually flowers only singly except when culturally suited. The purple-red form is illustrated in *The Specialist Orchid Grower*.

↑ *BULBOPHYLLUM GLOBULIFORME* (Nicholls). Perhaps the smallest orchid in any genus. The red pin head, upper centre, indicates the size of pseudo-bulb and flower. It is rarely found in its native land, Australia.

↓ *BULBOPHYLLUM GUTTULATUM* (Wallich ex Hooker f.). One of the most beautiful of the genus, rarely cultivated. It is from upper India, growing epiphytically at about 1000 metres (3250 feet). An intermediate climate orchid.

→ *CALANTHE ROSEA* (Lindley) Bentham. The pure white variety of this species. Burmese and Malaysian, the genus from the tropical to sub-tropical climates grows best in natural environments of similar rating.

↓ *CALANTHE VESTITA* (Lindley). The variety *turneri*, found in some of the orchid's habitats. A sub-tropical to tropical species, best grown to its potential in similar climates. Adapts to enclosures, but less floriferous and vigorous in cool areas.

↑ *CATASETUM* natural hybrid, obviously with one parent as *Catasetum pileatum* (Reichenbach f.). A natural step in the evolution of the genus, it has all the attributes of a species yet is not so. Probably *Catasetum splendens* (see *Orchid Review*, Vol. 3, p. 2).

↓ *CATTLEYA AMETHYSTOGLOSSA* (Linden and Reichenbach f.). Occasionally referred to as a variety of *C. guttata* (Lindley). Suited by intermediate climates. A Brazilian bifoliate.

↑ *CATASETUM HOOKERI* (Lindley). One of the more than seventy Brazilian members of the genus, and of the strangest group of orchids in the world catalogue. They carry male, female and sometimes hermaphrodite flowers, even on the one inflorescence.

→ *CATTLEYA AURANTIACA* (Bateman ex Lindley). A Central American bifoliate, usually suited to intermediate conditions.

↓ *CATTLEYA PERCIVALIANA* (O'Brien) var. *ALBA*. There are several pure white *Cattleya labiata* varieties, some rare, others not as plentiful as normal varieties but commonly found in early importations.

↑ *CATTLEYA WARNERI* (T. Moore). Frequently considered a form of *Cattleya labiata* (Lindley). A Brazilian orchid. Intermediate climate.

← *CATTLEYA WARSCEWICZII* (Reichenbach f.). Frequently listed as a Colombian variant of *C. labiata* (Lindley). It also has the synonym *Cattleya gigas* (Lindley and Andre). The *C. labiata* group is dealt with more fully in *Growing Orchids, Book Two*.

↑ *CERATOSTYLIS RUBRA* (Ames). A species which proliferates into a large plant, properly grown open-rooted and preferably in an intermediate to subtropical climate. Flower size about 2 to 3 centimetres (1 inch). Native to Philippine islands.

→ *CHILOSCHISTA USNEIOIDES* (Don) Lindley. A remarkable orchid, almost leafless; the root system acts for plant function. Native to northern India, preferably grown in an intermediate climate and slab-mounted.

↑ *CHONDRORHYNCA AROMATICA* (Reichenbach f.) P. Allen, syn. *Warscewicziella wendlandii*. A beautiful Central American epiphyte which is also known by other names. These species are without pseudo-bulbs and must be cultivated carefully to prevent fungoid diseases from lodging in the basal bracts. This species may be referred to as a cochleanthes.

↓ *CHYSIS BRACTESCENS* (Lindley). Most frequently quoted as a species rather than a variety of *C. aurea* (Lindley). The flower is from the immature green growth behind it. The older pseudo-bulbs are leafless, a stage of the annual cycle. They are cast in autumn, the new growth and flowers appearing in spring. Central American, grown best in intermediate environment. The genus was dealt with in *Growing Orchids, Book Three — Vandas, Dendrobiums and Others*.

→ *CIRRHAEA DEPENDENS* (Loddiges) Reichenbach f. A Brazilian epiphytic species, oriented towards an intermediate climate.

→ *CLOWESIA ROSEA* (Lindley) Reichenbach f. Allied to catasetums and separated from that genus because of its flowers, which carry pollen as well as seed capsules or ovaries. A warm-growing epiphyte suited to baskets or slat containers. For further information, see *Growing Orchids, Book Three*.

→ *COCHLEANTHES DISCOLOR* (Lindley) Schultes and Garay, syn. *Chondrorhynca discolor* (Rafinesque). A Central American species from an intermediate to warm climate.

↑ *COCHLIODA ROSEA* (Lindley) Bentham. A Peruvian species, allied to *C. noetzliana* (Reichenbach f.) Rolfe. An intermediate climate species, frequently difficult in the wrong environments. Flowering season spring to early summer. Information on the genus is included in *Growing Orchids, Book Two — The Cattleyas and Other Epiphytes*.

← *COELOGYNE CRISTATA* (Lindley). Rightly termed one of the easiest orchids to grow, but found difficult to flower. A cool-climate species with a liking for about 60 to 75 per cent shade. The change-over from growth to flowering is quick and much depends on ripening the pseudobulbs in late summer to autumn in brighter light. See *Growing Orchids, Book Three — Vandas, Dendrobiums and Others*.

↑ *COMPARETTIA MACROPLECTRON* (Reichenbach and Triana). A Colombian epiphyte from intermediate climates. Difficult to keep virile in artificial environments; stable and progressive in natural yet protected conditions ranging from about 15 to 18 degrees (60–65°F.) and mildly humid in warmer seasons.

→ *CONSTANTIA CIPOENSIS* (Porto and Brade). A small species from the harsh flatlands of Brazil and growing only on vellozia, to the flaky exterior stem of which this propagation is still attached. The flower is about 1.5 centimentres (⅝ inch) across. Dry-growing species, suited by cool to intermediate climates, appears tolerant.

← *CORYANTHES MACULATA* (Hooker). A Central and South American genus. This species common to many areas. A reproduction from an earlier publication. The genus is difficult to import and grow, and the best chance is with locally raised seed. A warm climate orchid, possible in intermediate environments.

↑ *CRYPTOCHILUS SANGUINEA* (Wallich). A newly imported plant flowering on the new growth. This is a poor specimen, but illustrates how some genera withstand change-over in climates and still grow and flower. An Indian orchid suited by cool to light intermediate conditions.

← *CUITLAUZINA PENDULA* (La Llave and Lexarza), syn. *Odontoglossum citrosmum*. A Mexican species from a cool climate, very easy to grow. It resembles oncidium in habit, hard leafed, the flower spikes coming from immature new growth in spring. Humidity and watering important, because flowers spot easily once opened.

→ *CYCNOCHES PENTADACTYLON* (Lindley). A warm-growing Brazilian and Peruvian species. The habit of growth is obvious, the pseudobulb well developed, lower sheathing bracts dried off and the upper leaves to be discarded as flowering finishes. Infrequent watering and low humidity during dormancy are essential. See *Growing Orchids, Book Three* for further information about the genus.

↓ *CYNORKIS RIDLEYI*. This genus is related to habenarias and is in some ways also reminiscent of calanthes. The flower stem is about 20 centimetres (8 inches) tall, the flowers not long lasting.

↑ *CYPRIPEDIUM JAPONICUM* (Thunberg). As a cultivated orchid the genus is neither popular nor easy. If fresh plants are obtainable rather than importations the species are much easier to grow. Terrestrials, needing study and understanding. Cool.

↓ *DENDROBIUM AGGREGATUM* (Roxburgh). The somewhat rare pale form of this Asian species. Recommended for intermediate environment and preferably slab-mounted. Flowering period, usually in summer, follows a dry dormancy through winter. Confusion persists about the difference between this and *D. jenkinsii* (Wallich ex Lindley).

↑ *DENDROBIUM AMETHYSTOGLOSSUM* (Reichenbach f.). Philippines species, not widespread, growing at about 1400 metres (4500 feet).

→ *DENDROBIUM BELLATULUM* (Rolfe). A member of the nigro-hirsute section and never long-lived for most cultivators. Cultural problems associated with these are not solved by cultivation in mixed generic groups, but a challenge hard to resist. Intermediate climate.

↑ *DENDROBIUM BULLENIANUM* (Reichenbach f.). Epiphytic, flowers in spring, mostly on defoliated stems. Flowers about 8 to 10 millimetres (up to ½ inch) across, produced in umbels or clusters. Philippine species, fairly widespread.

↓ *DENDROBIUM CAPILLIPIES* (Reichenbach f.). A miniature species which, although the flowering is sparse, is among the more beautiful of Thai species. The habitats spread across Burma and into northern India.

↑ *DENDROBIUM DELICATULUM* (Kraenzlin). A miniature species from New Guinea, the pseudo-bulbs about 1 centimetre (3/8 inch) long, flowering profuse, deep purple-blue and in the leaf-like pseudo-bulbs almost indistinguishable. Cool-growing, and from an altitude of about 1000 metres (3250 feet), it flowers in mid-winter. May be wrongly termed warm-growing.

→ *DENDROBIUM SECUNDUM* (Blume) Lindley. An Asian species native to Thailand and Borneo, the pendant racemes densely flowered. Warm-growing, suitable for intermediate to warm climates but probably best suited in the tropics. Spring flowering in cultivation.

← *DENDROBIUM SENILE* (Parish and Reichenbach f.). Adaptable to sub-tropical cultivation. Spring flowering.

↓ *DENDROBIUM SUBCLAUSUM* (Rolfe). Native to the Moluccas, to the west of Irian Jaya and between that land mass and the Celebes. The form of dendrobium is unusual and represents one of the few westerly species of the section calyptrochilus. It is warm-growing, although occurring at a fairly high altitude. First flowered by Veitch in 1894 and described in the Kew *Bulletin*.

↑ *DENDROBIUM TRIGONOPUS* (Reichenbach f.). Depending on origin in its Asian habitats, has variable tolerance and flowering period. Prefers intermediate to warm conditions and probably flowers in spring in cultivation.

↓ *DENDROCHILUM GLUMACEUM* (Lindley), syn. *Platyclinis glumacea*. A Philippines species which, although not as floriferous as *D. cobbianum* (Reichenbach f.), nevertheless has free-flowering habits and very few vices. The pseudo-bulbs are close set and need repotting infrequently. Principally spring flowering.

← *DIPTERANTHUS PUSTULATUS* (Kraenzlin). Brazilian temperate climate epiphyte, preferably grown open-rooted. Small species such as this need individual cultivation to keep plants healthy and flowering. Originally on fibre, secondary mount processed cork.

↓ *DISA* (Bergius). A mixed group of the genus, species and hybrid forms. Cultivated as a concentrated exercise by G. Sobey at Castlemaine, Victoria, Australia, and most successfully.

↑ *DISA CARDINALIS* (Linder). Brightly coloured but smaller than *D. uniflora* (Bergius).

↓ *DISA UNIFLORA* (Bergius). A well-coloured form of this beautiful disa. It may disappoint when grown with the usual collection of species orchids from many countries.

↑ *DISA TRIPETALOIDES* (Linnaeus f.). Infrequently grown except in collections devoted to the genus or in its native territory. In common with *D. uniflora* (Bergius) the correct climate, cool and moist, is one of the essential components in a program.

← *DORITIS PULCHERRIMA* (Lindley). With similar morphology to phalaenopsis in plant form and flower, it added much to colour factors in hybrids. Culturally treated as a phalaenopsis and needing warm environment but tolerant of less humidity.

↓ *DRYADELLA EDWALLII* (Cogniaux). Once known as masdevallia, this small species has three closely related Brazilian orchids, all difficult to distinguish. Easy to grow in cool to lightly intermediate environments. Spring flowering.

→ *ENCYCLIA BACULUS* (Reichenbach f.) Dressler and Pollard. The peculiarity of the inflorescence, with back-to-back flowers, is notable in this group; also the labellum is uppermost. Native to South America then after a gap across several states to Mexico. Like others, it has several synonyms.

↓ *ENCYCLIA CITRINA* (La Llave and Lexarza) Dressler. One of the most beautiful of the Mexican species, growing in a cool climate and capable of such culture. One of the more difficult orchids to grow and seldom long-lived in most collections. Spring to summer flowering.

↑ *ENCYCLIA MARIAE* (Ames) Hoehne. Probably merits equality with *E. citrina*. Also solely Mexican, is capable of cool growing but needs understanding for its dormancy. Spring to summer flowering, but variable according to environment and climate.

↓ *EPIDENDRUM DIFFORME* (Jacquin). A Mexican species which despite its soft coloured flower, grows sometimes in harsh climates. A reed-stem type, common to much of the country and extends into the southern part of United States of America and many bordering countries to South America.

↑ *EPIDENDRUM MEDUSAE* (Lindley) Bentham. One of the more curious orchids native to Ecuador. Grows in intermediate environment, flowers in spring to summer. Inconsistent.

↓ *EPIDENDRUM STENOPETALUM* (Hooker). A short-growing reed-stem species which carries few flowers on the apex of growths. Common to much of Central America, the Caribbean islands and South America. Flowers in spring and occasionally in late winter.

↑ *EPIGENIUM COELOGYNE* (Reichenbach f.) Summerhayes. Once known as a sarcopodium. The habit is morphologically long, the gaps between pseudo-bulbs making open-root mounts necessary. Flowers are borne singly, appearing spring to summer. Native to warm Malaysian climate, needing intermediate conditions and with marked dormancy.

← *ERIA CORONARIA* (Lindley) Reichenbach f. A member of a diverse group, once known by several synonyms. This species common to much of northern India, Burma and across to Thailand and Malaysia. Cool to intermediate, preferably intermediate environment. Flowers summer to autumn.

→ *ERIA JAVANICA* (Swartz) Blume, syn. *Tainia stellata* (Pfitzer). In some literature still referred to by its synonym. The difference in the two species illustrated is marked. Both are sub-tropical orchids and need intermediate culture, preferably in temperatures which do not fall below 12 degrees (54°F.).

↑ *EULOPHIA GUINEENSIS* (Lindley). Tropical African terrestrial. Flower spikes are tall and appear after summer growth. Climate intermediate to sub-tropical; would probably succeed in warm conditions. Shade is important for eulophias.

→ *EULOPHIA KEITHII* (Ridley). The eulophias comprise a large genus; this one is a terrestrial orchid from Asia. Seldom cultivated, perhaps because of origin as terrestrials, which are not popular with epiphyte growers. Occasionally quite tall, the flowering spikes must be kept well below the roof of enclosures. Flowers in late summer to autumn.

↑ *GALEANDRA DEVONIANA* (Lindley). Occasionally tall-growing epiphyte. The flowers appear in late summer, which dictates fast growth commencing in late spring. Temperate climate, or sub-tropical conditions with high humidity in growth stage. This species from Guyana, Brazil and Colombia, but the genus also Peruvian.

← *GONGORA GALEATA* (Lindley) Reichenbach f. A Mexican orchid and strange, like other members of the genus, The flower spikes pendant, the individual stems arching out from the main inflorescence, with the flowers opening toward this main stem. Intermediate climate, light shade; the flowers open in summer.

→ *GONGORA GRATULABUNDA*. A species about which nothing is known, but exhibiting the peculiarities of the genus.

→ *GRAMMANGIS ELLISII* (Lindley) Reichenbach f. A Madagascan species which should be grown in a warm climate but not tropically humid for long periods. An epiphyte with pseudo-bulbs, this plant was flowered in Melbourne in a moderately warm glasshouse, blooming in late summer to autumn.

↑ *GRAMMATOPHYLLUM SCRIPTUM* (Linnaeus) Blume. Native to parts of Asia and across to the Philippine islands and Niugini. It flowers in late spring and summer but may be variable in cultivation. A tropical species, better left to that climate rather than grown in a glasshouse in a cool region.

↓ *GROBYA AMHERSTIAE* (Lindley). A strange but beautiful species from Brazil. Intermediate to cool climate, the potting material based on cymbidium mix. Flowers in summer on immature new growths.

→ *HABENARIA RODOCHEILA* (Hance). One of the smaller habenarias. These orchids occur world-wide but do not comprise a great proportion of cultivated species. *H. rodocheila* is native to Thailand and other adjacent states. Intermediate to cool growing terrestrial, not cold.

↓ *HARAELLA RETROCALLA* (Kudo). A small Taiwanese epiphyte allied to vandaceous orchids such as saccolabiums. Such diminutives are apt to become lost in large collections. To prevent this they should be open-root specimens, hung where they get adequate shade. Intermediate to warm climates and lightly humid.

↑ *HEXISIA BIDENTATA* (Lindley). A small but brightly coloured epiphyte from the tropical area of Central America. Preferably an open-root, slab-mounted plant, it multiplies into a sizeable specimen. Warm cultivation, light shade.

↓ *HUNTLEYA MELEAGRIS* (Lindley). A rarely found species from Central and South America. Needs warm treatment, but not tropical. Overhead watering, which tends to fill leaf axils, inadvisable.

↑ *JUMELLEA GRACILIPES* (Schlechter). A Madagascan species with only one representative known in South Africa. The plant is similar to angraecums but has only single flowers from the leaf axils. The monopodial stem branches from the base; it is usually epiphytic and occasionally lithophytic. For culture, see the sections on angraecoid orchids (Angraecums, Aerangis, etc.).

→ *LAELIA ALAORII* (Brieger and Bicalho). From Bahia area, north-eastern Brazil, and a lowland climate. An epiphyte suitable for pots or slab mounts. Intermediate to moderately cool environment.

← *LAELIA BAHIENSIS* (Schlechter). From Bahia, in moderate environment, and lithophytic as frequently as epiphytic. Intermediate climate.

↓ *LAELIA BRIEGERI* (Blume), from the Minas Geraes area, Brazil, and needing an intermediate to warm climate, not subtropical. Lithophytic species, variable colour.

↑ *LAELIA JONGHEANA* (Reichenbach f.), from Minas Geraes. Cultivation similar to *L. dayana*. Minas Geraes is north of Rio de Janeiro, which is on the east coast of Brazil.

↓ *LAELIA LOBATA* (Lindley) Veitch. Native of Brazil and growing in a restricted habitat. This laelia is supposed to have confused Gardner in his discovery of *Cattleya labiata*, a very easy mistake to make considering its single-leafed pseudo-bulbs and cattleya-like flowers usually appearing in late spring to summer.

← *LAELIA PUMILA* (Hooker) Reichenbach f. A cool-growing species from widespread habitats and includes in its varieties cool-growing forms and those needing intermediate environments. Spring to autumn flowering.

↓ *LAELIA SINCORANA* (Schlechter). Frequently bracketed with *L. pumila*, but originating from the Bahia region, where *L. pumila* is not known to have occurred. Cool-growing, and tolerates cold: this plant has flowered in temperatures down to 6 degrees (43°F.) in winter in my experience.

→ *LAELIA SPECIOSA* (Humboldt, Bonpland and Kunth) Schlechter. A Mexican species, named as cool-growing but preferring intermediate environments. Flowers in summer to autumn, on new growth.

↓ *LEMBOGLOSSUM BICTONIENSE* (Bateman) Halbinger, syn. *Odontoglossum bictoniense*. A transferee from its genus, it flowers in autumn. A Mexican and Central American orchid suitable for cool climates. Among other species transferred, *Odonto. cervantesii* (La Llave and Lexarza).

← *LEMBOGLOSSUM CORDATUM* (Lindley) Halbinger. Cool-growing, common to Mexico and much of Central America, originating in cool forest at about 2000 metres (6500 feet). Spring to early summer flowering.

↓ *LEMBOGLOSSUM ROSSII* (Lindley) Halbinger, syn. *Odontoglossum rossii*. A winter-flowering species which must be treated with respect in other seasons. Listed as cool-growing, it succeeds in relatively low temperatures and does not like hot summers. Humid coolness in this season aids growth and prepares the plant for flowering. Mexican and Central American.

↑ *LEPTOTES BICOLOR* (Lindley). An intermediate climate Brazilian species easily grown into a large clump. It is preferably potted but will take to fern mounts. A watering style must be worked out to promote growth or dormancy in the appropriate seasons.

↓ *LUDISIA DISCOLOR* (A. Richard) Lindley. Not noted for its flowers, this beautiful orchid is prized for its foliage. Two distinct varieties are illustrated. Native to Burma and other sub-tropical regions, intermediate environments are advised.

← *LUDISIA DISCOLOR* flowers are about 15 to 20 millimetres (up to ¾ inch) across, almost pure white, appearing in spring to early summer on new shoots. The growths elongate to about 10 or 12 centimetres (4 to 5 inches). It is climatically variable.

↓ *LYCASTE SKINNERI* (Bateman ex Lindley). The accepted name at present, but likely to be changed again. This species exhibits many diverse forms but this is probably among the best. An orchid from cool climates, it needs intermediate environments in artificial cultivation. Known also by its synonym *Lycaste virginalis* (Scheidweiler).

→ *MACRADENIA RUBESCENS* (Barbosa Rodrigues). A Brazilian species from a small genus growing in Amazonas and Para. *M. rubescens* is extinct in one of these states and doubtful in the other because of the destruction of the habitat. A species with latitude into intermediate environments.

↓ *MASDEVALLIA BURFORDIENSE* (O'Brien). A Peruvian species for intermediate to cool conditions. Flowers in autumn. It needs constant humidity throughout the year of 50 to 60 per cent.

↑ *MASDEVALLIA CONCINNA* (Koniger). South American species. Cool to intermediate climate, autumn to winter flowering.

← *MASDEVALLIA DISCOLOR*. A member of the widespread genus native to the Central and South American states. Spring flowering, it is one of the miniatures morphologically, but has large flowers.

↑ *MASDEVALLIA GLANDULOSA* (Koniger). South American species. Intermediate climate, flowers spring to summer.

↓ *MASDEVALLIA IGNEA* (Reichenbach f.). One of the brightest of species, native of Colombia, cool-growing but will not tolerate temperatures lower than 10 to 12 degrees (50–54°F.). The upper sepal always drops over the flower like a hood. Spring to summer flowering, variable.

↑ *MASDEVALLIA STROBELLII* (Sweet) Garay. One of the most beautiful of the genus. Ecuadoran species, intermediate environment, but tolerates cooler conditions.

← *MASDEVALLIA UNIFLORA* (Ruiz and Pavon) 1794. One of the earliest of the genus to be named. This species varies from white to rose. Native to Peru and Ecuador.

→ *MASDEVALLIA VEITCHIANA* (Reichenbach f.). Frequently considered the outstanding member of the genus. Variable in colour and also flowering period. Cool-growing.

↓ *MAXILLARIA LEPIDOTA* (Lindley). A Colombian, Ecuadoran and Venezuelan species, showing the unusual flower potential of the genus.

↑ *MAXILLARIA NIGRESCENS* (Lindley). Venezuelan species, cool-growing. Tolerates temperatures below 10 degrees (50°F.) but prefers minimums above that. Flowers in autumn to winter, strongly perfumed.

← *MAXILLARIA SANDERIANA* (Reichenbach f.). The largest of the genus. This plant is flowering from spikes through the base of container. Flowers are brilliant when first opened, but the red loses intensity after a few days. Ecuadoran species; flowers spring to summer, but most variable.

↑ *MAXILLARIA SEIDELII* (Pabst). Brazilian species, among the small members. Grows cool, flowers about early spring to early summer. Native of Rio de Janeiro state. A miniature.

↓ *MEIRACYLLIUM WENDLANDII* (Reichenbach f.). A small species from the Central American region, needs intermediate to warm climate, flowers in autumn. These small plants create much interest in collections, particularly those devoted to species.

← *MILTONIA BLUNTII* (Reichenbach f.). A natural hybrid but qualifies as a species orchid. One parent has been transferred to another genus, which in reality makes *M. bluntii* an inter-generic hybrid. Flowers late summer in cool climates. A Brazilian orchid.

← *MILTONIA SPECTABILIS* (Lindley). A Brazilian species, morphological counterpart of the miltoniopsis, the Colombian branch of the genus. Flowering season usually summer, variable, the spike appearing on immature new growth. Cool cultivation, but thrives also in intermediate environments.

→ *MILTONIOIDES WARSCEWICZII* (Reichenbach f.) Brieger and Lueckel. A Central American and South American species transferred from miltonia to miltoniopsis and now to the genus miltonioides. This is the variety *alba*. Considering the history and relating it to other genera similarly transferred, it is little wonder growers are confused. The authority quoted here is Leon Wiard's *Introduction to the Orchids of Mexico* (1987).

→ *MILTONIOIDES CARINIFERUM* (Reichenbach f.) Brieger and Lueckel. The generic title is queried. This species has been titled oncidium and odontoglossum in its cultivated history. Central American and South American orchid suited by intermediate environments and also possible in cool conditions, grown as an odontoglossum.

↑ *MILTONIOPSIS PHALAENOPSIS* (Linden and Reichenbach f.) Garay and Dunsterville. Known in isolated specimens from Central America, but credited largely as a Colombian orchid. Intermediate climate, with miltonia culture; warmer, but similar to odontoglossums.

← *MILTONIOPSIS ROEZLII* (Reichenbach f.) Godefroy-Leboeuf. The pure colour or *alba* form, Colombian and other states origin, should be grown in intermediate climate in suitable conditions.

↑ *MILTONIOPSIS VEXILLARIA* (Reichenbach f.) Godefroy-Leboeuf. The pure colour or *alba* variety of this orchid, rare like *M. roezlii*. A Colombian species suited by intermediate climate.

→ *NAGELIELLA PURPUREA* (L. O. Williams). A cool-growing species, flowers late summer and autumn. Flowers freely but the blooms are small and a large plant should be cultivated.

← *NEOBENTHAMIA GRACILIS* (Rolfe). Height of the flowering stems and the bulk of the plant makes it more suitable for outdoor growing in a sub-tropical climate. Intermediate temperatures. The individual flowers are about 1 to 1.5 centimetres (½ inch); the head is about the same size as reed-stemmed epidendrums.

← *NEOFINETIA FALCATA* (Hu). A Japanese orchid with resemblance to the angraecum complex of Madagascar. Small and delicate, it is supposed to be cold-growing, but does better in cool to intermediate climate. Flowers autumn to spring, occasionally early summer in some locations.

→ *NEOLEHMANNIA PEPEROMIA* (Reichenbach f.) Kraenzlin, syn. *Epidendrum porpax* or *Neolehmannia porpax*. Common to many centres in South America and Central America. Considered cool-growing, but much would depend on origin of plants. Brazilian section found at about 1000 metres (3250 feet), Amazonas.

↓ *NEOMOOREA IRRORATA* (Rolfe). A Colombian species, the flowers are about 6 centimetres (2½ inches) across. It has no colour variations, and the plant is about the size of the largest lycaste.

← *NOTYLIA BARKERI* (Lindley). A Central American warm-growing species. Infrequently cultivated and flowering in summer.

↑ *OBERONIA PALMICOLA* (F. Mueller). Growing naturally in northern Queensland in the fork of a tree. Not frequently cultivated, and little interest is shown in the genus. Unspectacular but carries many small flowers, autumn to spring.

← *OBERONIA MUELLERANA* (Schlechter). Usually flowered with a vertical inflorescence and with colour variation. Individual flowers are about 2 millimetres (⅛ inch) diameter or a little more.

↑ *ODONTOGLOSSUM CIRRHOSUM* (Lindley). Native to Colombia and Peru. From high altitude, cool-growing, flowering winter to spring. Efforts to grow it in cool climatic zones are usually unrewarding.

↓ *ODONTOGLOSSUM CRISPUM* (Lindley). Probably the best-known of the genus. Colour variation in the species remarkable, including brown-red and yellow-gold blotched and clear, principally in irregular patterns. Flowers winter to spring, cool-growing, Colombian; this form is known as the Pacho type.

← *ODONTOGLOSSUM HALLII* (Lindley). A Colombian and Ecuadoran species tentatively stated also as Peruvian. Grows at about 2500 metres (8200 feet). Cool-growing.

↓ *ODONTOGLOSSUM HARRYANUM* (Reichenbach f.). A Colombian and Peruvian species, imperfectly expanding at times. Cool-growing, from about 2000 metres (6500 feet). Possibly also Ecuadoran.

↑ *ODONTOGLOSSUM STELLATUM* (Lindley). Native of Guatemala and credited as Mexican, possibly also Costa Rica. Lower intermediate to cool, probably best suited in cool environment.

↓ *ODONTOGLOSSUM TRILOBUM* (Schlechter). A Peruvian species, not discovered until 1921, growing at about 1200 to 1500 metres (4000 to 5000 feet). Suited to cool culture (see page ix).

↑ *ONCIDIUM EXCAVATUM* (Lindley). A cool-growing species from Peru and Ecuador, flowers midwinter. This orchid will grow and flower in temperatures as low as 6 degrees (43°F.), the blooms on branching racemes and lasting for weeks. It can be grown in pots or on slabs, preferably the latter because of its climbing habit.

← *ONCIDIUM FORBESII* (Hooker). A Brazilian warm-climate species, adaptable to intermediate environments. Flowers autumn to spring. Habitats in Amazonas, Sao Paulo, Rio de Janeiro and Espirito Santo. Variable colouring, area to area.

↑ *ONCIDIUM ORNITHORHYNCHUM* (Humboldt, Bonpland and Kunth). Central American species consistent throughout the habitats; flowering season winter to early spring. Intermediate climate orchid, best grown in natural environments rather than enclosures because of flowering season and growth habits.

↓ *ONCIDIUM PAPILIO* (Lindley) growing with phalaenanthe dendrobiums in a natural climate under fibre-glass roof, North Queensland. An intermediate to warm species, flowering winter to spring. Native to South American northern states and Trinidad. Usually unsuccessful in cool climates, however artificially warmed: the flower stems are shorter and flowering disappointing.

← *ONCIDIUM ZAPPII* (Pabst). A Brazilian oncidium for intermediate climate, growing in bright conditions. Native to only one area, Espirito Santo.

↓ *ORNITHOCEPHALUS IRIDIFOLIUS* (Reichenbach f.). A bundle of leaves and roots which produces flower in summer to autumn, this small species adds to numbers where that is important. For the ardent species collector it represents beauty equal to any other. The slab is what is termed granulated cork of coarse texture.

→ *OSMOGLOSSUM PULCHELLUM* (Bateman ex Lindley) Schlechter. A beautiful winter-flowering species native to Central America. From a cool climate, it is among the hardy orchids which flower in temperatures of 6 to 8 degrees (42–45°F.). There may be up to eleven flowers to the wiry stem; this one is 3 centimetres (1¼ inches) across. For other information, see *The Specialist Orchid Grower* or *Growing Orchids, Book Two — The Cattleyas and Other Epiphytes*.

↓ *PABSTIA JUGOSA* (Lindley) Garay. A Brazilian species from the southern states of that country. Intermediate climate, adaptable to warm environment. Not subtropical or recommended for hot areas. Spring to summer flowering.

↑ *PALUMBINA CANDIDA* (Lindley) Reichenbach f. A species which would be notable in any collection. It is native to Guatemala and, supposedly, Mexico. This plant was in my collection about 1960 but I disposed of it and have not seen it since. Intermediate climate, not cool or hot. Related to oncidiums.

← *PANISIA UNIFLORA* (Lindley) Lindley. This orchid, native to northern India, Nepal and the Khasia Hills at 250 to 1500 metres (800 to 5000 feet) is supposed to belong to the genus coelogyne. It is considered suitable for intermediate climates.

→ *PAPHIOPEDILUM BARBIGERUM* (Tang and Wang). A species of southern China. Discovered or rediscovered in 1940. In the books *Growing Orchids* most of the species were illustrated. These are new, prefer intermediate culture and are best left out of cool collections.

→ *PAPHIOPEDILUM DIANTHUM* (Tang and Wang). Allied to *Paphio. parishii*, this southern Chinese species exhibits differences from it. A lithophyte from limestone country.

← *PAPHIOPEDILUM HAINANENSIS* (Fowlie). A species from Hainan Island. Belonging to the *P. appletonianum* group, it appears to have insufficient distinction for separation from that section.

PAPHIOPEDILUM HENRYANUM (Braem). A variable orchid recently added to the list of newly discovered species.

↑ *PAPHIOPEDILUM MALIPOENSE* (Chen and Tsi). A warm-growing species of southern China, spring to summer flowering. Plant with tesselated or marbled foliage, propagates from extensions of rhizome.

→ *PAPHIOPEDILUM MICRANTHUM* (Tang and Wang). A southern China species first collected in southern China in 1940, described in 1951. Intermediate climate Slipper. Flowering times in natural habitats seldom reflected in artificial culture. Usual southern hemisphere period is winter.

↑ *PAPHIOPEDILUM WARDII* (Summerhayes). Discovered in western China at about 1350 metres (4500 feet). There is a considerable list of species from this region, all of which should be considered unsuitable for cool to cold cultivation in the temperatures quoted on page ix.

← *PERISTERIA ELATA* (Hooker). A tropical orchid from the Central American region, extending to South America according to some authorities. Grows to about 1.5 metres (5 feet). The head of the spike is tender and must be adequately shaded. Flowers are about 4 to 5 centimetres (1½ to 2 inches). Flowering period summer to autumn. Cultivation, warm.

↑ *PHALAENOPSIS LUEDDEMANNIANA* (Reichenbach f.). A Philippines species most variable in colour. Some forms are pale cream, barred with red-brown markings, others brownish with light patches. A warm-climate species, the flowers are about 4 centimetres (1½ inches) across.

↓ *PHALAENOPSIS MANNII* (Reichenbach f.). A northern Indian species flowering in summer and a warm-growing orchid. Small and for some a purely botanical specimen.

↑ *PHALAENOPSIS SCHILLERIANA* (Reichenbach f.). An early colour photograph of a well-flowered plant, indicating the capabilities of the species. The plant was imported from Sanders', England, before 1939. Philippine islands species.

↓ *PHALAENOPSIS VIOLACEA* (Teijsman and Binn). Native to Malaya, Sumatra, Borneo. Carries few flowers, variable colouring from country to country. It is a tropical orchid, and needs protection from burning, like all the genus. Flowers in late spring, but variable.

↑ *PHRAGMIPEDIUM LONGIFOLIUM* (Reichenbach f.) Rolfe. South American species from Minas Geraes, Brazil. Over this wide area it occurs as a minor species, principally in high country with high humidity. Most of the genus are shade-loving, demanding about 60 to 70 per cent shade cloth, darker in hot sunny weather. Intermediate to warm.

→ *PHRAGMIPEDIUM SARGENTIANUM* (Rolfe). Centred on the Slipper family, these orchids are similar to the Asian members but remain distinct. This orchid is native to South America and requires warmth. The leaves are quite long and the flower stems carry consecutive blooms for some months in winter to spring.

↑ *PHRAGMIPEDIUM SCHLIMII* (Reichenbach f.) Rolfe. Short-petalled species from Colombia and Peru, occasionally drab coloured. Like all the genus, they are rarely epiphytic, mostly terrestrial or lithophytic. Use potting material similar to that for paphiopedilums and preferably leave undisturbed for long periods. Warm-growing to intermediate.

← *PHREATIA* species, unknown. Principally native to the islands north of Australia and up to the Philippines. This plant came from Niugini at the foot of the Kokoda Trail. It is typical of the genus.

↑ *PLEIONE FORMOSANA* (Hay). An Indo-Asian genus with many different forms bearing this title. They are generally among the orchids which grow in cold conditions but preferably above frost level. Flowers precede growth on new shoots in spring. A number of new species have recently been added to the catalogue.

↓ *PLEUROTHALLIS PTEROPHERA* (Cogniaux). A Brazilian species, small and easily grown into a specimen plant. Widespread occurrence in southern Brazil from Rio de Janeiro southward. One of the largest genera, many of which are yet to be found and named.

← *POLYSTACHYA PUBESCENS* (Lindley) Reichenbach f. A southern African terrestrial orchid, member of a numerous genus in that country. Autumn to spring flowering and prefers cool environment.

↑ *PLEUROTHALLIS TRIBULOIDES* (Swartz) Lindley. A Central American species, morphologically small, the flowers, clustered at the base of the leaves, brick red. Intermediate climate. Conditions which are too cool will halt growth and stultify flowering in spring to summer.

← *PROMENAEA OVATILOBA* (Klinge) Cogniaux. Similar to *P. xanthina*, but not widespread and occurring only in Espirito Santo area of Brazil. Cool climate epiphyte, spring to autumn flowering.

↑ *PROMENAEA STAPELIOIDES* (Link and Otto) Lindley. A Brazilian epiphyte, occasional lithophyte, from southern Brazil. The foliage is tender on this genus and should be kept dry while growing. Flowers spring to summer on new growths. Cool environment.

→ *PROMENAEA XANTHINA* (Lindley). Most widespread of the species and quite variable in form. Colour consistent, sometimes more yellow. Foliage a little harder than that of *P. stapelioides*. Epiphyte, very occasionally lithophytic, flowers spring to summer, cool environment.

↑ *RENANTHERA MONACHICA* (Ames). An epiphyte from the lowland, hot regions of the Philippines, the genus spreads over the northern part of the archipelago. In cultivation a tropical climate with bright light is a necessity.

← *RESTREPIA HEMSLEYANA*. Many of the genus from South America not positively identified and some classed as pleurothallis. This is a larger member and most attractively coloured. Intermediate cultivation, either in pots or mounted on treefern. A Colombian species.

↑ *RHYNCOLAELIA DIGBYANA* (Lindley) Schlechter, variety *fimbripetala*. Slightly yellower than the type and with fringed petals. This Honduras species needs a warm climate. Difficult in environments with high annual temperatures, occasionally makes continual growth instead of maturing and then flowering in summer to autumn.

→ *RHYNCOSTYLIS COELESTIS* (Reichenbach f.). A Thai species needing intermediate to warm environment. Flowering late autumn and early winter.

← *RHYNCOSTYLIS GIGANTEA* (Lindley) Ridley. Extensive areas of occurrence through Thailand to Burma. Epiphytic, needing intermediate to warm climate. Flowering late winter and spring, but variable in cultivation, according to local environment. Slotted pots or slat baskets are ideal containers.

↓ *ROBIQUETTIA CERINA* (Reichenbach f.), syn. *R. merrillii* Ames. A Philippines epiphyte, the flower head is typical of the crowded inflorescence of most of the genus. Foliage and habit similar to rhyncostylis. Warm environment, lightly tropical.

↑ *RODRIGUEZIELLA GOMEZOIDES* (Barbosa Rodrigues) Pabst, syn. *Theodorea gomezoides*. Southern Brazilian orchid needing intermediate to slightly cool conditions. An epiphyte. The genus contains about five Brazilian species.

↓ *ROSSIOGLOSSUM GRANDE* (Lindley). A Central American species which has had constant admiration over its history. Cool to intermediate environment, late summer to early autumn flowering.

SACCOLABIUM LONGIFOLIUM (Lindley), syn. *Acampe rigida*. The last name is probably the correct title. This is one of the small 'botanical interest' orchids. A widely disseminated species through the Indian sub-continent, intermediate environment, spring flowering.

← *SCUTICARIA HADWENII* (Lindley) Hooker. A Brazilian orchid which is infrequently cultivated. Needs a warm climate but probably adaptable to intermediate environment under good control. There are five species native to Brazil and their identification is confused.

↑ *SOBRALIA XANTHOLEUCA* (hort. ex Williams). An epiphytic orchid, occasionally lithophytic, native to Guatemala. It grows naturally at about 1500 metres (5000 feet) in a warm climate. In cultivation it should be suited by intermediate to slightly warm environments. It is not a tropical type.

↓ *SOPHRONITELLA VIOLACEA* (Lindley) Schlechter. The sole member of the genus, Brazilian native, suitable for cool-growing environment. Although it may take time to develop a large plant, there is little degradation in the process, the foliage lasting a long time.

← *SOPHRONITIS ACUNAE*. A species supposed to originate in Bolivia which on examination differs little from *S. cernua*. It appears to grow in slightly cooler conditions.

↓ *SOPHRONITIS BREVIPEDUNCULATA* (Cogniaux) Fowlie. This plant flowers midway between *S. mantiqueirae* and *S. coccinea*. This last-named orchid is given three flowering periods in Fowlie's analysis, indicating the disarray caused by ignoring the fact that natural flowering may differ from that of plants grown in artificial environments. I believe that the genus must be actively grown for more than ten years to understand it. The cult may be better served if the genus was consolidated like the *Cattleya labiata* complex into varieties instead of species.

→ *SOPHRONITIS CERNUA* (Lindley). A warmer growing orchid than *S. coccinea*, this Brazilian species is not widely cultivated. It is an epiphyte, occasionally lithophytic, best grown open-rooted.

↓ *SOPHRONITIS MANTIQUEIRAE* (Fowlie). Although Fowlie has tried to instil some verity into his dissection, in practical growing the species tend to flower regularly but not in the sequence proposed, nor have I been able to translate the plant descriptions to fit my collection (see page 249).

↑ *SPATHOGLOTTIS PLICATA* (Blume). An Indo-Asian, Philippines and Australasian genus, all but one or two needing warm climates. Most species grow best in the sub-tropics, where they may be cultivated as garden plants. In this situation they make their own arrangements and are not stifled by containers. *S. plicata* grows almost world-wide as, at times, a wild orchid. It originated in the Malaysian area.

← *SPIRANTHES NAVARENSIS* (Ames) L. Williams. A terrestrial species of the Panama region. Grows a rosette of leaves upward, the spike of flowers coming from the centre about late winter to spring. Warm environment but adaptable to intermediate. A large genus of some 300 members, almost world-wide distribution.

→ *STANHOPEA OCULATA* (Loddiges) Lindley. One of the orchid genera which send their flower spikes down through the potting material. They are decorative but must live a suspended life. This inflorescence broke the rules and emerged through the side. Some species are reputed to be cold-growing, but prefer cool to intermediate environments.

↓ *STELIS PAPAQUERENSIS* (Reichenbach f.). The genus is large and at times ill-defined. The flowers are similar almost throughout the species and although small are most beautiful in detail. This species is native to Brazil, Guyana and Venezuela. It needs an intermediate environment.

↑ *STENOCORYNE SECUNDA* (Vell.) Hoehne. A Brazilian species, should be given intermediate to warm climate but not sub-tropical.

← *SYMPHYGLOSSUM SANGUINEUM* (Reichenbach f.) Schlechter. Peruvian orchid, preferably cultivated in intermediate conditions, although occasionally referred to as growing in odontoglossum conditions. Basic requirements the same, but minimum temperature should be higher.

↑ *THUNIA MARSHALLIANA* (Reichenbach f.). Bamboo-like terrestrial orchid from the Burma area, preferably cultivated in intermediate environment. Growth should be quick, followed by flowers in late summer. Growth renewed annually, flowering only once.

→ *TRICHOGLOTTIS BRACHIATA* (Ames). Member of a Philippine islands genus, vanda-like in habit and growing up to 40 to 50 centimetres (16 to 18 inches) tall. Flowers in late spring to summer, but can vary when transferred between continents or hemispheres. Warm climate.

← *TRICHOGLOTTIS WENZELII* (Ames). Philippines species, widespread through the islands, grows to more than 1 metre (up to 4 feet) tall. The genus branches frequently and is prolific in producing side shoots in the correct climates. *T. wenzelii* needs warm conditions and head room in enclosures.

↓ *TRICHOPILIA MARGINATA* (Reichenbach f.). Central American genus which many growers find difficult to maintain. Seasonal habits must be understood and warm to intermediate environments provided. Natural habitat about 1000 to 2000 metres (3250 to 6500 feet), but in tropical climate. Epiphytic.

↑ *TRISATELLA HUEBNERI* (Schlechter), syn. *Masdevallia huebneri*. Some authorities spell it as triaristella. Native of Amazonas, Brazil, intermediate climate probably best, but with restrictions on upper limit temperature. See the section on masdevallias for cultural information. Like dryadellas, the species were transferred from masdevallias.

↓ *VANDA COERULEA* (Griffith). Burmese to Thailand origin, flowers in autumn to winter. Preferably upper limits of cool conditions, although this plant experiences quite cold nights in its habitat.

↑ *VANDA COERULESCENS* (Griffith). Some could regard this as the poor Burmese relative of *V. coerulea*, but it has much to recommend it. Still cool to intermediate growing, this epiphytic plant appears easier to grow than *V. coerulea*. It flowers in winter.

↓ *EUANTHE SANDERIANA* (Reichenbach f.) Schlechter. The paramount species as well as the only one. Even if left in its original genus as a vanda it would still be paramount. A Philippines orchid, it requires hot conditions rather than warm, although provided the warm, humid climate is constant the plant is very rewarding. An epiphyte and only very occasionally lithophytic, a branching monopodial, *Vanda* or *Euanthe sanderiana* will multiply with side shoots, but is slow.

↑ *VANDA WATSONII* (Rolfe). This species is almost indistinguishable from *Vanda kimballiana* (Reichenbach f.). A vanda carrying terete leaves with a deep channel. This Asian and Burmese duo — if we regard the two as synonymous — grows in intermediate climates and is not sub-tropical or tropical. The flowers are sparse and the plant multiplies by branching from side shoots.

→ *VANDOPSIS GIGANTEA* (Lindley) Pfitzer. Probably the largest of the genus, a species of Burma and Thailand, it prefers warm to hot climates. It also occurs with coarse, thick foliage.

← *VANDOPSIS PARISHII* (Reichenbach f.) Schlechter. The habitats of this species extend from upper Burma to Thailand. The foliage is coarse, short and burns easily in light which is too bright. The environment should be upper intermediate to warm and there is scope for growing the plant in hot but not tropical environments. This a single flower of the variety *marriotiana*, which has short spikes.

↓ *ZYGOPETALUM CRINITUM* (Loddiges). Has been found epiphytic as well as terrestrial; flowers on new growths in spring. Usually lighter coloured and shorter spiked than *Z. mackayi*, carrying only five or six flowers. Lightly perfumed.

↑ *ZYGOPETALUM MACKAYI* (Hooker). One of the easiest orchids to grow in cool climates. A terrestrial and unusual at that, this species flowers on the new growth in autumn and matures its pseudo-bulbs over the winter and early spring, most frequently starting new growth before the flowered bulbs are mature. Brazilian species, heavily scented on bright sunny days.

→ *ZYGOSTATES LUNATA* (Lindley). A widespread Brazilian species, best suited by intermediate to warm environment. Botanical interest to growers of miscellaneous orchid genera, free flowering in late summer to autumn.

Meiracyllium (Reichenbach f.) 1864

Most species orchid growers have one or more miniature species in their collections and they are a source of interest to those privileged to see them in flower. Every area of the world has many of these delightful small orchids and there are growers who prefer them to the larger and more showy species.

Meiracyllium is one for those fortunate enough to be able to get orchids common to the Central American and Mexican region; most agree that it is more difficult to get orchids from those countries than from the rest of the American continent. It is a small genus of about four members, all of which are plants made up simply of leaves and rhizomes, like some of the pleurothallis and restrepias.

M. wendlandii (Reichenbach f.) grows in most areas common to the genus. It is small and brightly coloured; the leaves take on a reddish tinge when grown in the correct light, which gives them as much beauty as the flowers. Some authorities give only one other species to accompany *M. wendlandii*, while others add two more. The principal part of the two-species group is *M. trinasutum* (Reichenbach f.). More interest is being taken in these small orchids and many are being added to the total catalogue each year.

The history of these orchids can be found in the records of plants exhibited at Royal Horticultural Society meetings, and a plant of *Meiracyllium gemmae* was exhibited at one of these. While it could be researched further and equated with either of the two modern cultivated species, the meeting was held in 1896. The name *M. gemma* is a synonym for one of the modern titles of the two.

The natural habitats of these small orchids mark them as bright-light species from a warm climate. The habit of growth indicates the type of cultivation they need. The plants should be attached to strips of cork or other acceptable material and suspended in bright conditions where they get adequate sunlight for most of the day.

These miniatures are apt to be forgotten, so the plants should be placed where they will not be overlooked for watering, flowering or dormancy.

Small genera such as meiracylliums should also be regularly fertilised when noted as growing or flowering, particularly when buds are swelling and opening. They have little or no pseudo-bulb capacity, and correct fertilising increases the size of flowers and maturing of the plants.

Miltonia (Lindley) 1837

This genus was named so early in the history of cultivation that, as one would expect, some species have been transferred into newly created genera. In general the basic name miltonia has been retained, with suffixes to contain those orchids which did not conform to the initial requirements for flowers in the genus.

The morphology of the total group is remarkably similar and allied as it is to odontoglossums and oncidiums the new section miltoniopsis usually contains those species with flowers having more exaggerated labellums and other features.

The miltonias are principally Brazilian orchids with rather harder morphological characteristics than the Colombian miltoniopsis. The flowering habit is different also. Where miltoniopsis have few flowers to the raceme, miltonias may carry quite long racemes and more flowers.

As outlined in *Growing Orchids, Book Two* and in the later volume *The Specialist Orchid Grower*, the genus is full of anomalies and misplaced species and it is best for growers who have no way of ascertaining corrected names to keep the labels on their plants, however old or out of date those names may be. In time the system will right itself, but taxonomists and would-be taxonomists do a disservice to the pastime if they try to impress too much their convictions of what is right or wrong. It should be noted that *Sander's List of Orchid Hybrids* is a useful source of information, particularly the lists of the years 1981–5.

While some species could be considered intermediate class which will grow and flower in temperatures as low as 10 to 12 degrees (50–55°F.) and lower, some do better if they are given a few degrees above that and have minimums which do not fall below 12 degrees (55°F.). Others again will grow and flower, usually a little later in the season, in temperatures as low as 6 to 8 degrees (42–48°F.). The different species should be chosen for their capacity to withstand the climate in which they are to be grown, and passed over to someone else if the plants do not thrive within a couple of years.

Most Brazilian miltonias grow much better if they are tied to slabs of treefern which give the plants incentive to grow upward. The rhizomes of most are similar to those of oncidiums and follow the climbing pattern of those orchids.

In watering plants mounted on treefern slabs it is better to take the fern slab off the wall or wherever it hangs and water it thoroughly from the back rather than the front. Fertiliser should also be applied in this manner. In good environments, do not water or fertilise unless the plants are flowering or growing.

Miltonioides (Brieger and Lueckel) 1983

New and confusing, this genus is made up of members which were at one time known as odontoglossums, oncidiums, miltonias and other genera. The morphological distinctions were such that they needed attention. But again it is possible that the names may not be adopted, principally because of lack of communication and publicity amongst the orchid-growing community.

In general these orchids grow in intermediate temperatures and in environments slightly dryer than those usual for miltonias or odontoglossums. Flowering periods vary according to the species, but could be noted as summer or autumn until they prove otherwise. The prevailing habit will be as in odontoglossums or miltonias, with plants growing in the warmer months of the year and flowering on completion and fattening out of the newest growths.

Habitats of various species are widespread, in Mexico, through other Central American states and into the South American states of Colombia, Peru and Ecuador. In those countries there is no direct relationship other than through taxonomic and botanical analysis, but right through the genus there is similarity.

There is no relativity in cultivation of the new species with the miltonias of Brazil and few have the climbing habit or rhizome of those orchids. Pots and potting material such as used for the finer-rooted epiphytes should be tried first, and if found unsuitable the lead should be taken from what the plant indicates, such as assuming a climbing habit.

The growing season is the time to note the habits of various species. Unlike odontoglossums, which have the unusual habit of growing perhaps at what seem unseasonable times, most miltonioides will fall into regular patterns.

Watering and fertilising irregularly growing plants produces a problem: the whole collection cannot be treated to, say, a Sunday morning task after a busy week. It is tempting to use a hose and trust to luck that the right thing has been done, but results will not be good. Again the double-potting method is recommended (see page 81). There is little need to argue about the condition of that inside potting material: the outside surface of the clay inner pot tells the story every time.

Generally the fertilisers which are successful for odontoglossums or miltonias will be found suitable, but if above-average results are looked for in special subjects a little extra trouble pays off. The origins of transfers to miltonioides will be found in the illustrations on page 162.

Miltoniopsis (Godefroy-Lebeuf) 1889

A name which originated in the late years of the nineteenth century, but which was not applied to the genus until almost one hundred years later. It refers to that section, mostly Colombian, which had to be separated because of morphological differences from other plants included under the title miltonia.

The miltoniopsis and derivation of the name were mentioned in *Growing Orchids, Book Two*. The principal members or species of the genus are *M. vexillaria* (Reichenbach f.) and *M. roezlii* (Reichenbach f.), both of which are principally Colombian orchids. *M. roezlii* is also native to areas of Central America, but isolated only and not general.

There is a substantial difference in the cultural patterns of the genus compared with the miltonias of Brazil and miltonioides. The foliage is softer, with less substance and prone to develop spotting or die-back of the tips in conditions too cool or too hot. This softness is also apparent in the pseudo-bulbs and the plants generally.

Temperature range for good growth and flowering is within a narrow band between 15 sand 28 degrees (60–80°F.). Although not a good part of the cultural program, what is known as a hot-box is the ideal situation for the plants over the colder months in temperate to sub-temperate climates. This piece of equipment is usually designed to run on low-voltage heating strips or cables and is controlled by thermostat. The required minimum temperature is set, and the heating cable is activated automatically when necessary to warm the bed of material through radiation. The system gives control within about 5 degrees (10°F.), supplying miltoniopsis with the type of climate in which they thrive naturally. Natural climates are unusual in cultivation.

The system had its counterpart in what were known as 'stoves' in the early history of orchid growing and before the value of air flow was realised. The plants were immersed in the contents of the box or stove.

This genus requires warmth combined with a constant flow of air, so when it is grown in an enclosure there must be a means of drawing air in, with a warming element coming into operation whenever the internal temperature falls below the minimum. The hot-box operates intermittently throughout the day and night when wanted. The air flow, whether warmed or of outside temperature, must be constant. With this system the author was able to grow miltoniopsis, hybrids and species, to realise their potential. The moisture content of potting mixes can be controlled by wetting the contents of the hot-box, without applying water directly to the plants; which should be constantly damp. Liquid fertilisers should be applied as for other genera.

The species *M. vexillaria* and *M. roezlii* show little variation from clone to clone and this led to confusion in the middle years of the twentieth century, when hybrid registrations were suspended for some time. The first primary hybrid, *Miltonia* Bleuana, was raised in 1889. A great amount of interbreeding between it and the two species resulted in the inbreeding of the genus because of paucity of fresh species material.

In *Reichenbachia*, Frederick Sander said: 'This magnificent orchid is a hybrid. [*M.* Bleuana.]... Both parents are indigenous to the United States of Columbia [sic] and both are equally beautiful and continue to be grown in every collection....' It is interesting to note that in conformity with primary hybrid vigour in almost all cross-pollinations, the flowers were much larger than either species.

Other species, of course, are available at times and these should be carefully grown, because the orchids of Colombia and contiguous states have been reduced to a small fraction of the numbers once noted.

Potting material should be fairly fine. I used the finest grade of the best bark available, which aggregated about 2 millimetres (1⁄16 inch) in diameter, with no additives at all. The pots were small and crocked to a small fraction of their depth. The species *M. roezlii* (Reichenbach f.) Nichols grew and produced two or three flowers for a number of years in an 8-centimetre (3-inch) pot. Taken at face value, while hybrid miltoniopsis may grow into larger plants which produce a quantity of flowers, with at times two spikes from each pseudo-bulb, the risk with species is that the root system, fine as it is, may not be able to fill out in larger pots and will deteriorate more quickly.

For the genus it is difficult to predict flowering times. In the usual cultivation and weather patterns of temperate climates, the plants are dormant with partly made growths slowly growing; they reach flowering stage in late spring and the flowers open in summer.

There is always temptation to try to force growth and flowers, which, if they come at the wrong time, are more likely to contract disease and fail to mature. It is frequently imagined that hybrid orchids follow patterns laid down by species, but this is untrue because of the mixture of species involved, which in the late years of the twentieth century are more than the original two used.

If miltoniopsis are grown as additions to hybrid orchid collections, they should not be tried as single plants because of the demands they make on time and patience.

Nageliella (L. O. Williams) 1940

The two species of the genus originally were known as *Hartwegia purpurea* and *H. angustifolia*, the first-named known since the 1830s. There is some misinformation about the genus generally, and the species *N. angustifolia* is occasionally referred to as a variety of *N. purpurea*. In general the genus prefers intermediate climates and light shade.

The genus is Central American and Mexican. Some would call them untidy looking plants, but they grow freely in the right conditions and well-suited specimens often flower continuously. The flowers are small and bright; on well-grown plants the old stems continue to bloom until they are completely worn out and dried off.

In any collection of orchids the species *N. purpurea* is an ornament. It should be tied to a piece of cork, hardwood which has been well weathered, or a section of treefern or fibre.

An early reference to *Hartwegia purpurea* describes the plant flowering at the Kew Gardens, with a note on the beautiful variegated foliage and its comparative rarity in cultivation. This rarity is reflected in present-day orchid growing, with the species almost lost in the flood of other species.

Four other species are listed in different sources of information: *Hartwegia kienastii* (Reichenbach f.), who placed it under the synonym *Ponera kienastii*; *H. genna* (Reichenbach f.); and two others which should never have been included in the genus.

It is interesting to note also a brief reference to the genus: 'It can only be called a botanical orchid, but it is interesting to find that it is still in cultivation and to learn something more about it.'

Nageliella purpurea as a naturally occurring plant grows at altitudes from 1000 metres to about 2000 metres (3250 to 6500 feet), usually as an epiphyte growing in half-light and occasionally in quite adverse-looking situations where the leaves are subject to altered colour and the flowering sparse.

In cultivation the usual watering pattern for orchids should be followed, with frequent applications in the growing period, as the plants have little substance to store moisture and disseminate it in dry periods. The plants do not shrivel much in such situations, but there is so little substance to them that, taken with the rather straggling appearance, it is not altogether noticeable.

However, care should be taken not to overwater the plants when they appear dormant, even if they are flowering. In general the growth period occurs during warmer months and flowering occurs at almost the same time. The usual liquid fertilisers applied in this period will suit.

Neobenthamia (Rolfe) 1891

The genus comprises a single species which is native to the tropical regions of East Africa. It bears no relationship to the benthamias of the Madagascar region. The prefix 'neo' means new and the two genera were a tribute to Bentham, a noted botanist and taxonomist who added much to knowledge concerning orchids in the nineteenth century.

Neobenthamia gracilis (Rolfe), the species once known as a member of the polystachya group, is remarkably similar to the reed-stemmed epidendrums in habit, with a head of flowers at the apex of the slender stems that reach heights of 1 metre (3 feet) and more. There is no pseudo-bulb system and in this they also resemble the arundinas and other orchids which are almost solely terrestrial.

The flowers are about 2 centimetres (½ inch) across, invariably white. They open successively as the head elongates; usually it becomes so heavy that the stem bends at the top. The flower head tends to be horizontal instead of remaining vertical like *Epidendrum ibaguense* or others with similar morphology and stiffer stems.

As a cultivated orchid *N. gracilis* has never been a popular species; perhaps it was because of its white flowers that it never impressed growers. Had the flower head been red it would probably have been sought for its decorative value. There is little relief from the white other than minor areas of purple-red and yellow.

Sanders included the genus in their *Orchid Guide* and gave Zanzibar as the source of their material. Plants first came into cultivation about 1884, but the species was not flowered until about 1890. Growing naturally the plants are far more straggly than epidendrums and the habit persists with growers. More suited to sub-tropical climates, it is a difficult subject for glasshouse culture, because it occupies a lot of space and the flowers are not impressive. However, it flowers easily and if cultivated in small clumps of up to a dozen canes it is worth a place.

Light is an important part of culture of *N. gracilis* and shade should be little more than about 50 per cent. If grown without enough light, the plants appear to be unable to ripen the stem to the point of flowering. Once having borne their blooms the stems remain, but unless new growth is formed at the base of older and flowered growths there is no continuity.

Cultural conditions, if grown in the right climate, would be similar to those for arundinas or thunias, aimed at fast growth. Occasional shoots from upper areas of older growths may also grow and flower, but they add to the ragged appearance of clumps.

Once the plants are growing and flowering the watering should be regular but not lavish. Fertilisers, solid or liquid, make for better flowering.

Neofinetia (Hu) 1925

This delightful plant may be compared with vandas and angraecums. It has a long history of cultivation and is not noted in species orchid collections as often as it merits.

Originally known by several synonyms, it was first flowered in the collection of the Horticultural Society in England and named as *Angraecum falcatum* in the records of that society in 1819. But that apparently was not its first appearance: an English grower named Hume cultivated and flowered it some years before that. The plant, the only species in the genus, is native to Japan and Korea. It is considered a cold-growing orchid, but is more accurately described as cool-growing. Shade should be moderate, of 60 to 75 per cent.

The difference between cool-growing and cold-growing is vast, even though the temperature varies by only three or four degrees. Cool growing must be based on no lower than 10 degrees (50°F.) and preferably 12 (54°F.). It is regrettable that growers do not take sufficient notice of these restrictions and avoid the tendency to relate 'cool' to their particular environment.

N. falcata will grow and flower quite well in temperatures above the figures quoted but will not thrive in the warmer regions where summer maximums are constant and in the area of 25 to 30 degrees (77–95°F.). The resemblance to some of the African and Madagascan orchids is not confined to the flowers, as the general habit of the plant is that of an open-root epiphyte which succeeds best when grown on a slab rather than in a pot.

It is not the only stranger with the look of an angraecum found in the western Pacific region, as the species *Amesiella philippinense* (Ames) Garay, from the Philippine islands, also has the long spur and relatively distinct appearance of those plants.

In good cultivation the pure white flowers should not be splashed when misting or watering, as they soon spot. Humidity in the glasshouse, where these plants are best suited, should not be too high, as this also has the effect of spotting them. Not long lasting, they should remain good for about a fortnight to three weeks.

There is a growth phase and this is followed by flowering, usually in autumn to early spring, but the period may vary according to culture. The roots should adhere to anything used as a mount; they grow on cork or any well-weathered piece of wood. The root system should be profuse, with much remaining aerial and unattached.

Neolehmannia (Kranzlin) 1899

There is some confusion in designating species as belonging to this genus and various publications give different versions. One of the best-known of the species apparently rightly belonging to it is *Neolehmannia porpax* (Reichenbach f.), which is also known by the synonyms *Epidendrum porpax* (Reichenbach f.), *E. matthewsii* (Reichenbach f.) and *E. porphyrophyllum* (Schlechter). To confuse things still further, several other synonyms are apparently ignored.

The genus is native to South America and areas of Central America, where this species is still known as *Epidendrum porpax*. Its right name could possibly be *Neolehmannia peperomia* (Reichenbach f.), the title given in the illustrations, which is preferred by Pabst and Dungs.

Whatever name is subsequently chosen for this beautiful miniature orchid, most growers appreciate its charm and the ease with which it may be grown. There are other species in the genus, some close morphologically, the difference being in the labellum, which is the principal and most apparent part of the flower. *Venezuelan Orchids*, Dunsterville and Garay, Vol. 6, hints at synonymy.

In cultivation it could be expected that the plants would be ideally suited to slab or fern block culture, but because the root system is sparse and short and the plants inclined to be succulent, they seldom thrive when grown this way.

Each plant in total is a mass of short growths rather than pseudo-bulbs, distichously leafed and flowering from the head of the growths with short stems. Once having flowered, the stem may produce more than one offshoot; in a few years if properly suited it becomes a tightly packed mass of live and dead stems.

It is easily judged if culture is right because poorly grown plants and those grown in too bright light have dark-coloured leaves, short stunted growths and small flowers.

Shade should be of at least 60 to 75 per cent. The plants may be grown in flat dishes or shallow pots in whatever mixture suits the cultivator. The mixture must not be dank or stale, and it should be moist throughout the year. Apply fertiliser only when the plant is flowering or in growth. Usually new side-shoots appear on the older stems in the warmer periods of the year, yet the plant is not a warm-growing genus. Flowers appear usually in spring to early summer on mature stems.

It may appear strange to allot space to so small a genus as the neolehmannias, but when the plants are properly grown and carry a number of their strange, bee-like flowers they repay the interest.

Neomoorea (Rolfe) 1912

A Colombian genus of possibly only one species, although some suggest there are two. It is one of the rarest of the semi-terrestrial species and uncommon in cultivation although almost one hundred years have passed since its first notice.

Originally known as *Moorea irrorata*, the generic name was pre-empted by another plant and so became invalid. The history is related in the *Orchid Review* in its early volumes.

Apparently the plant from which the first flowering at Kew, England, emerged was a propagation of one of the first of the genus brought into cultivation and grown at Glasnevin, Dublin, a place renowned for orchid and plant culture. It was featured in the *Gardeners' Chronicle* for 1892. The plant was at first thought to be a lycaste until it flowered.

Neomoorea irrorata (Rolfe) is unusual for reasons other than its rarity, as a glance at the illustration will indicate. Its morphology is similar to the more robust of the lycastes and its culture follows closely the recommendations laid down for that genus. It has been known to produce between twenty and thirty flowers on its spikes, not all of which open together; in cultivation the number is usually reduced and may total only ten or twelve. Fertilise as for lycastes.

Propagation is by divisions and it is hoped that pollinators are able to work on this orchid with the purpose of producing seedlings. It is one of the worthwhile projects which may take years to fulfil because of scarcity of clones. In such as this species the best system is to have two separate clones with even moderate differences in the flowers to work on, as frequently self-pollinations do not set seed.

Potting for these orchids follows those for lycastes, phaius, calanthes and similar broad-leafed semi-terrestrial orchids. The use of the term semi-terrestrial is advisable because frequently these plants are found in almost epiphytic states. The potting medium should be such that the root system will not be affected by watering at the wrong time. The plants grow and flower almost in the same season, which is spring to early summer.

Although it could be imagined that such orchids would grow in sub-tropical conditions, it would be best to describe their needs as warm temperate; they do not favour cold or long periods of poor light such as could be expected in glasshouse culture in cool climates. It could be held in favour of neomooreas that they are probably dormant or partly so during the coldest part of the year and would suit some situations in cool temperate climates. Shade needs to be of 60 to 75 per cent, and the foliage should not be allowed to yellow or burn.

Notylia (Lindley) 1825

A genus which has aroused little interest, perhaps because the species, of which there are about forty, have not been readily available. The range of this genus reaches across the northern part of South America and enters Central America. In Brazil there are about twenty species and they grow in almost every state of that country and through into Bolivia.

Generally they are related to gongoras, catasetums, maxillarias and other similar sympodial orchids. Some species are of medium size, while others such as *Notylia norae* (Garay) are very tiny, measuring barely 2 centimetres (¾ inch) across the total leaf span. Such orchids when available create a demand which cannot be satisfied, despite the fact that the flowers are unspectacular. The miniature morphology attracts attention.

The genus is best adapted to slab mounting rather than potting by any method. Plants should be securely fastened to such things as slabs of cork, weathered hardwood or pieces of hard treefern rather than softer slices.

It is difficult to attach these tiny plants to their mounts, but try to fix them firmly. If the plant is loose, its roots will be abraded every time it is moved, and you will have to untie it and start again. Once the plant is secure, the roots are able to grow and make their own attachment. When this stage is reached the tie may be cut away.

Unsuitable mounts frequently cause trouble — if a mount is too acid or too alkaline the roots will refuse to attach and another type of mount must be substituted. Where cork is used it may be found that natural cork bark is unsuitable while granulated and compressed cork may suit better. This can be noted only when the growing season has started and new roots move out from the base of new growth.

The ties used for attaching plants are usually selected for lasting quality. Nylon fishing line of suitable size and thin strips cut from old nylon hosiery are among the best. Both these materials are springy and may be hard to knot, but several turns before pulling the knot tight usually makes the fibres bond and hold firmly.

Notylias are intermediate to tropical orchids and should not be grown in cool to cold areas. Water should be applied only when plants are growing, and shade should be light but not to the point of burn. Fertilisers should be sprayed on the plants, or they may be dipped once a week in the growing period.

Oberonia (Lindley) 1830

One of very few Afro-Asian genera, its habitats stretch from East Africa through India, Burma, the western Pacific islands such as the Philippines and down through Niugini to the Solomon islands and Australia. There are more than two hundred species.

Over this vast expanse of earth's surface there is little change in the plants' morphology. It is typified by the compressed habit of growth; the leaves, if they may be called that, are flattened and markedly distichous. Although the species may be difficult to name unless in flower, in their flowering the total genus is remarkably similar, with hundreds of minute flowers perfectly arranged on slender, mostly drooping racemes.

Not a widely cultivated genus, plants add something to collections devoted to culture of native orchids, wherever they may originate. The needs of the genus vary, but taken as a whole it may be regarded as suitable for intermediate to moderately warm conditions. There is no system of pseudo-bulbs, as most epiphytes have; the rather coarse substance of the leaves takes its place and new growth emanates from the base of older growths and what there is of a rhizome system. The roots are fine, and potting material should suit this type.

Plants of *Oberonia palmicola* (F. Mueller) love shade and will thrive in situations in which most epiphytic orchids would not set seed; they produce long spikes of tiny flowers, numbered in hundreds, which attract pollinators. Their seed can travel far on the wind: the author once found small seedling plants of *O. palmicola* growing at a distance of 3 kilometres (2 miles) from the nearest producing group of the species. This occurred in northern Queensland and across an expanse of canefields. Finding them was an enjoyable experience.

Plants in good cultivation develop multiple points, with most tending to produce a clump. These distichous growths elongate and flower for two or three years; then thay act as reservoirs of plant substance for the newer parts. Some flower only once.

In nature, the plants generally hang from trunks and branches. They have a drooping habit, and indeed the basal parts are too thin to support upright plants. The indications are for slab mounting, but note that the root system is fine and must adhere. When not growing or flowering they need little or no water. Most prefer a humid atmosphere, and fertilisers applied at appropriate times.

Odontoglossum (Humboldt, Bonpland and Kunth) 1815

Among the earliest cultivated genera, it originally included among its species some ten or more orchids which have been transferred to other genera or newly created genera. Some would say that of all orchids it is one of the most diverse and beautiful, originating in an area of the world comprising South American and some Central American states.

Odontoglossum species are described as cool or cold growing but the reality cannot be so generalised. Some species grow in cool conditions, as described on page ix. It is certain that the genus is best suited to cool climates, producing the best growth and flowers in those conditions.

Those who grow the species retained in the genus in the late years of the twentieth century have some difficulty getting stock, and the genus does not like the system of quarantining and treatment used in most countries. Some of the stock, particularly of the species *Odm. crispum* (Lindley), is remnant material from the 'golden days' of the genus about the end of the nineteenth century, when thousands of plants were taken from habitats in South and Central America. They were lost because of the wasteful procedure of keeping the best and discarding those which did not measure up to the desired shape or colour.

Odm. crispum is but one of many species sought to fill out collections devoted to cultivating native forms rather than hybrids. In this growers should be aware of differences which may be needed in cultivation methods — different temperatures, humidity and occasional dormancy of some of these species. Most of this arises from differing altitudes and climates at various points of origin of the plant.

Potting material and pots also enter the calculations. While some growers are content with plastic pots and find they suit their systems of growing odontoglossums, it is wise to take note of the advantages of the double-potting system described on page 81. Plants grown in double pots need only be checked once a week or so, and the amount of moisture can be finely adjusted to suit each plant's individual requirements.

The differences in culture are not confined to species such as *Odm. crispum*, but extend to those plants which have been transferred into other genera. These, such as rossioglossum, osmoglossum and lemboglossum are dealt with under their new titles. Each are different from those of odontoglossums, and in broad terms each genus requires its own particular climate.

Odontoglossums should have a separate growing area allotted to them; if other orchids are grown with them, they should be chosen for their suitability for the climate which odontoglossums prefer. Sometimes this can be found out only by trial and error, selling those genera and species which do not conform to odontoglossum growing systems.

The maximum temperature for odontoglossums should be no higher than 25 degrees (80°F.), dropping markedly at night to a minimum of 7 to 12 degrees in summertime and 10 to 12 degrees in winter (45–54 and 50–54°F.). This temperature range is stricter than that usually thought appropriate for amateur cultivation techniques. The humidity should relate to the temperature, but should never be less than 50 per cent or more than 80 per cent. This takes some management, and may be impossible in sub-tropical regions. The choice of equipment to control temperature and humidity depends on the incentive of the grower and the costs of installation and operation.

If you can approximate these conditions, they may suit nearly all the species it is possible to aggregate in many years of ordering and waiting for them.

The style of benching should also be considered, with solid benchtops preferable to the open mesh type which suits epiphytic orchids. The space between the roof and the plants should be such that the coolest part of the enclosure is occupied by the genus and this will be found about 60 centimetres (2 feet) above the floor. Always have the plants easily accessible without having to move one to get to another.

The final points concern air flow, water, and fertilisers. The plants should receive a constant flow of air, but not so much that they dry out rapidly. They prefer cool air. Watering and fertilising are inseparable in growing odontoglossums. Each plant should be watered separately, and the hose not used either as a spray-over or watering method. If the foliage is kept dry there is less chance of plants contracting the rots which enter through leaf axils. Fertiliser should be applied almost constantly to take advantage of the habits of the plants. They seldom rest.

Oncidium (Swartz) 1800

A genus of great complexity, native to Central America, South America in its warmest parts and the Caribbean islands. Over this range the genus varies considerably in its form: some species have definite pseudo-bulbs, but in others the leaves almost take the place of these pseudo-bulbs. The genus comprises more than eight hundred species.

The climates over which the species are spread vary from warm and dry to moist and humid. The species exhibit in their morphology something of the nature of their preferred climates. Perhaps the best examples are *Oncidium ornithorhynchum* (Humboldt, Bonpland and Kunth), from the softer yet warm areas of Central America, to *Oncidium luridum* (Lindley), which extends from parts of Florida, in North America, through the Central American states into Brazil, Guyana and Peru. This last-named orchid has what could be termed remnant pseudo-bulbs topped by heavy textured leaves which take their place. It is found in hot and at times very dry regions, and its appearance contrasts with the softer foliage and fleshy pseudo-bulbs of *O. ornithorhynchum*. That the latter prefers shady conditions is indicated also by its pink flowers. Most oncidium species exhibit yellow and brown colours, which are usually the product of bright-light conditions. Some species are epiphytic, others lithophytic as well as epiphytic.

The growth habits of all are to proceed upward; in some the habit is accentuated to the degree that they are almost impossible to cultivate in containers. Where plants must be benched to allow reasonable space for suspended plants, it is possible to attach them to lengths of fibre such as treefern or any other preferred mount. Insert the bases into pots and fill them with heavy material to stabilise them. The plants can be trained and tied to the 'posts' and so controlled that they remain there for a number of years.

Some species, such as *Oncidium papilio* (Lindley), have a tight system of pseudo-bulbs topped by a hard-looking single leaf; by the nature of their growth and appearance they suggest exposure to bright light and at times harsh conditions. Such orchids are not easy to grow in enclosures with other genera like cattleyas, which require high humidity during growth, and warmth and humidity during dormancy. The high temperature and drying out for some part of the day which is necessary for *Oncidium papilio* is unsuitable for other genera. The appearance of *Oncidium papilio* and other similar species is never one of robust, fattened pseudo-bulbs, but rather a slightly ribbed look. They are warm-climate orchids and should be left to those places with this type of environment.

Cultivation in artificial conditions should seek fairly fast growth. This is relatively easy in climates which are warm to hot for six months or more and then stabilise to moderate temperatures in other periods. An artificial climate, as well as being adjusted for warmth and humidity, must also provide sufficient sunlight. Where conditions for much of the year are dull, oncidiums soon exhibit irregular growing patterns and do not go through a fairly fast maturing phase and follow with flower. Not all species flower over summer and autumn; some flower in spring, some in what we usually regard as winter in cooler climates.

Very few species produce mature growths and then go into dormancy before producing flowers. In some climates such species as *Oncidium papilio* produce flowers in succession from the single stems produced by the leading pseudo-bulbs of plants. Four or more flowers may be produced from this single stem. In poor cultivation the stems may not even mature, let alone produce flowers.

Each species should be selected from knowledge of success or failure of fellow orchid-growers. The term cool-growing may be quite misleading when applied to, say, *Oncidium varicosum* (Lindley), which grows over the summer and flowers in early autumn. If conditions are cool in the early summer, the growth phase may take too long and the pseudo-bulbs will not have matured enough by the time flowering should occur. This is the most important point to note in selecting any oncidium.

Forcing oncidiums into the growth stage of their cycle is far from easy, even with temperature adjustments. It appears that day length, sunlight and natural habits are the governing factors.

The plants show in their root tips that the growth phase is about to begin. This may happen even when the pseudo-bulbs appear dry and ribbed. With oncidiums grown in pots it is difficult to observe the state of the root system. Only warmer-growing types such as the large-leafed species and one or two of the pseudo-bulbed, light-leafed species are best served by this method.

Light requirements for the genus generally are rather high. In regions with bright sunlight and long days the genus needs about 60 to 75 per cent shade. In climates with relatively cool and changeable summer weather, try shade cloth of 50 per cent and increase the density if leaf burn occurs.

Spotting of foliage is common. It is usually caused by fungal infection of soft foliage in its development stage, which in turn is caused by spraying or overhead watering.

Fertilisers should be liquid and applied only during active growth and when flower stems are apparent and continue to open.

A genus as large as oncidium cannot be dismissed in a few lines. The sheer variety of species makes generalisation meaningless.

Most pseudo-bulb types such as *Oncidium excavatum* (Lindley), which grows and flowers in temperatures as low as 6 or 8 degrees (42–46°F.), have racemes which may hold more than sixty individual blooms. These racemes may branch, and the flowers may take some weeks to open and mature to the tips in cool conditions.

In the best of conditions little more could be expected with such numbers. But in such as the equitant group, which has almost no pseudo-bulbs and is markedly distichous, with short but thick and channelled leaves, the flower stems may be long in comparison to plant size. The flowers which develop on the tips of the spikes to ten or more, open almost simultaneously. These small oncidiums are native to the islands of the Caribbean group and confined almost solely to that area. The Venezuelan species *Oncidium pusillum* (Linnaeus) Reichenbach f., syn. *Psygmorchis pusilla*, represents the form on mainland South America, otherwise there are few.

The equitant group is such that in daytime temperatures of 30 to 35 degrees (86–95°F.) and never much below 26 to 28 degrees (77–80°F.) it grows and flowers almost continuously to perfection. In cooler climates it is a possibility but never as vigorous as in warmer regions. Such challenges are best left out of well-considered species orchid growing in temperate or cool climates.

As almost the whole genus could be classified as warm to heat loving it is obvious that artificial heating is needed to grow many species. In cooler climates where the temperature in the colder parts of the year may seldom reach 20 degrees (68°F.) in daytime and fall as low as 8 to 10 degrees at night (45–50°F.) there are problems to be solved.

The method of growing the genus remains constant over the total temperature range, with complete outdoor cultivation possible under shade cloth of about 50 to 60 per cent in warmer regions.

Descriptions of various areas where species are found naturally do not help the amateur grower unless altitude and overhead cover and several other factors are given. The broad-leafed species with pseudo-bulbs absent or almost so are used to strong light; the leaves become quite dark in such situations. When grown in glasshouse conditions such leaves change in new growth to mid-green; only in the best conditions will they be as long or stout as those of the habitat.

Ornithocephalus (Hooker) 1825

This small genus of Central American, South American and Caribbean orchids is what some growers term 'of botanical interest'. Of course, all orchids in species growers' collections should be so classified and regarded.

The confusion into which these small 'botanical interest' orchids may fall is indicated by the fact that *Ornithocephalus gladiatus* (Hooker) had nine synonyms. Part of this was caused by the widespread occurrence of the orchid, which originated in twelve Central American, South American and Caribbean sources. As many as twelve different collectors, perhaps more, sent specimens back to their bases and herbariums, where the material was processed by botanists and taxonomists, leaving a considerable amount of work for their inheritors.

In habit the plants are similar to equitant oncidiums because they are also distichous — the foliage emerges from the rhizome and alternates side to side, making what could be taken to resemble a fan. It is a habit common to many genera.

The genus was cultivated in the early years of the pastime in Britain, with *Ornithocephalus gladiatus* (Hooker) blooming in the Glasgow Botanic Gardens in 1825. It is also known by the synonym *O. inflexus* (Lindley), native to Panama, Central America and other states.

In the late years of the last century an exhibitor received a Botanical Certificate from the Royal Horticultural Society for the novelty *Ornithocephalus grandiflorus*, which indicates the relative values of the time. In the considerations of a century later a specimen plant at least would be needed to gain recognition of such a relatively unimportant species.

In cultivation the genus is better suited to slab mounting on cork or similar material, and is capable of producing several spikes of flower on a single growth. In good conditions with adequate temperature control the plant could be expected to multiply with offshoots from the rhizome as there are no pseudo-bulbs.

Seven and possibly more species grow in Brazil and in their habitats in Amazonas, Para and Goias they all come from similar climates; generally these are warm and humid, the plants thriving in subdued light and with adequate moisture in the annual cycle.

The form of the plant gives clues to its habitat. There are no pseudo-bulbs, therefore no moisture storage to tide the plants through dormancy. The leaves of the plants are thick, although small. These leaves should be watched so that they do not shrivel; when root growth is seen the plant will absorb liquid nutrients.

Pabstia (Garay) 1978

This comparatively recent addition to the taxonomic catalogue absorbed the genus previously known as colax; before inclusion in that genus some species were also known as maxillarias.

In Veitch's *Manual of Cultivated Orchids*, Lindley was recorded as having made transfers because of dissatisfaction with some of the floral features. Other taxonomists also toyed with the species from time to time, but finally Garay renamed it in honour of Guido Pabst, who had much to do with the orchids of Brazil, apparently the only country in which pabstias are found. They come principally from the southern states of Rio de Janeiro and Sao Paulo, but the genus is also possibly represented in Bolivia. It does not extend into Venezuela.

Related to zygopetalums, pabstias are epiphytic and inhabit the climates and environments in which the larger portion of orchid flora of Brazil are concentrated. This climate is principally found in the coastal foothills and elevations up to about 1000 metres (3250 feet). Five species occur in Brazil, none of which are frequently noted in cultivation. The generic name colax was derived from Greek and inferred parasitism, a role often allocated to orchids in early years of their cultivation.

The plants of *Pabstia jugosa* (Lindley) are small compared with their relatives zygopetalums, with new growth commencing in early spring and flower stems appearing in new growths. As these growths lengthen so do the flower stems, which carry from one to four blooms on well-cultivated plants.

An interesting feature of these flowers occurs if they are pollinated. Instead of the flower parts shrivelling after turning brown, they thicken and colour dark green. They are by some assumed to act as leaves in carrying chlorophyll to aid seed development. In character, pabstias are no more curious for this feature than many other stages of orchid development and evolution.

Pabstias are usually grown in pots. The potting material should be relatively coarse, considering the size of the plants. Dryness should be a very minor part of the annual cycle, with non-growing and non-flowering stages the only time when moisture should be limited. Their natural habitat is in lightly humid forest regions. The potting material should never be saturated, but rather continually moist; do not allow the material to become stale or compacted. Danger time is in dormancy, when the root system is not assisting aeration.

Fertilising should be only moderate and infrequent during the growth period until plants mature; then it should be stopped and the plants allowed to enter their dormant period over late autumn and winter.

Palumbina (Reichenbach f.) 1861

A Guatemalan genus with a single species, the name is derived from the appearance of an alighting pigeon, which with a little imagination is clear in the flower of *Palumbina candida* (Lindley) Reichenbach f. The affiliation is deemed to be with oncidiums, but although the plant is similar the flower is totally different. The single feature which could be relevant is the labellum, the principal part of the flower. Overall colour is very pale pink.

The pseudo-bulbs are oncidium-like, with a single apical leaf which may grow to about 30 centimetres (12 inches) and is similar to leaves of *Osmoglossum pulchellum* (Bateman ex Lindley). Once established the plant is vigorous and quickly multiplies. It grows well and flowers in intermediate conditions, but in moderate to warm conditions soon loses vitality. Neither will it grow in cool to cold environments.

The collector Hartweg, when gathering plants for the Royal Horticultural Society in Guatemala in the 1840s is credited with having discovered the orchid. It has been rarely noted since that time and only once have I seen it in flower, a photograph of which appears on page 175.

Loddiges was apparently the inheritor of the survivors of that first collection which flowered. When his collection was broken up, a plant went to the Schiller collection in Germany, whence Reichenbach had his specimen. The remainder apparently died.

Lindley described the species *P. candida* in a summary from a drawing. He never saw it in flower. He noted:

> All the evidence I possess regarding this consists of a sketch from the live plants, made in 1843. Specimens have been lost and the growing individual disappeared. . . . If the sketch before me is accurate this cannot be an oncidium, but until the plant shall have been re-examined it is safest to leave it where it stands.

The flower is unusual because the two lower sepals are joined and hidden by the labellum. Although it appears to lack parts, it is complete. The racemes or spikes may carry five or more flowers, but most illustrations, photographic or lithographic, show only two or three flowers.

Orchids of Guatemala, by Ames and Correll, published in 1985, states that this orchid is rare and has not been collected for more than one hundred years. Such orchids are at risk: if one owns such a specimen it should be pollinated and seed raised. In general it is a plant for pots; the potting material should be fairly coarse. Cultivation is based on careful oncidium or maxillaria systems. Shading should be moderate, about 60 to 75 per cent.

Panisea (Lindley) 1841

The inclusion of somewhat rare orchids from small genera makes collections individual and remarkable. The genus panisea has only four members and of these only two come within the ambit of most amateur orchid growers. The two are, according to Alex Hawkes, *Panisea tricallosa* (Rolfe) and *P. uniflora* (Lindley).

Other sources, among them *Indian Orchids: A Guide to Identification and Culture* by Pradhan, give other names. Pradhan lists *Panisea uniflora* (Lindley) and *P. demissa* (D. Don). However, in a footnote he gives the true designation in his opinion of *P. uniflora* (Lindley) as a coelogyne, with the other one the only member of panisea. Those paragraphs give something of the confusion into which orchid growing can descend. It is impossible to correct names offhand and whatever the labels say should be accepted.

Paniseas inhabit many areas of northern India and extend into Burma. The regions include principally cool to intermediate climatic zones and the plants adopt epiphytic habit. They resemble coelogynes morphologically and their culture should follow a pattern of slab mounting rather than potting. Each system has its devotees, however, although more growers perhaps prefer less hanging material and more benched pots.

Both plants flower from the base of mature pseudo-bulbs. *P. demissa* has a longer raceme than *P. uniflora*, which, as the name indicates, has single flowers on short stems.

Altitude plays a part in selecting environments in which orchids should preferably grow, but in the instance of paniseas should not affect a choice of intermediate climate. If a choice had to be made for the coolest growing species it would be for *P. demissa*.

The author of *Indian Orchids* considers both species may be grown in shallow pans such as is recommended for pleiones and for growing *P. demissa* into a specimen plant. With more flowers to the raceme this recommendation is worth considering, however long it takes to produce a specimen plant. Both species are only moderate-sized plants and both appear to have the propensity to cast leaves readily. In order to grow specimen plants from such habits, it is always necessary to sever the older pseudo-bulbs and encourage new growth from the older parts rather than let them reach the stage where there are probably no viable 'eyes' left at the base of the growths.

Both or all species would have definite rest periods when not growing or flowering and shade should be moderate but not too heavy.

Paphiopedilum (Pfitzer) 1886

One of the medium-size genera, the species are almost all native to the northern hemisphere. It has been cultivated so long that the original plants grown were taken into captivity more than 150 years ago. Much space in publications has been given to the genus and it is not proposed to add much here.

The plants belong to the cypripediloideae. There are three principals in that family — the paphiopedilums, cypripediums and phragmipediums. The morphology is distinct and unmistakable. In general the flower lacks one part of the usual six segments comprising the flower parts of any genus. The ventral sepal, situated behind the labellum or pouch, is really two parts joined together. Occasionally this ventral sepal appears in split form. In assessing the flower on discovery of this, most opinions rank the flower as being deformed. The reality, of course, is that it is apparently a reversion in evolution to the original flower form.

The genus is said to be moist-growing and this is partly true. However, there is a difference between artificial cultivation and habitat, where during certain periods of the annual cycle the plants may dehydrate and remain dormant for some weeks.

In artificial cultivation a new set of conditions must be devised, particularly for growing in mass as most of the genus is cultivated. A program must be worked out to give the plants a medium set of conditions where slight dormancy is encouraged and allowed for in watering systems, without permitting dehydration.

In growing the genus, one must remember that the species originate in so many different environments that it is working against nature to try to grow all of them in the one system. Distinguishing characteristics require as many as three separate environments to accommodate half the known number of species. Most of the distinctions are produced by the different centres of origin, the altitude at which the plants grow and their positions relative to the ecology. The genus is totally shade-loving.

The number of species appears to increase each year, but the validity regarding some as species in their own right could well be questioned. Likewise, some original members of the genus appear to have modern varietal conceptions which do not fit characteristics. Taxonomy, too, is a point of concern, where the 'splitters' work their arts and make life difficult to understand. While the motives may be pure, there could well be suspicion when results are analysed.

In considering this section of the cypripediloideae there are two principal distinctions in the foliage and many more applying to the type of plant.

The foliage may be loosely divided into two forms — the plain green-leafed and the tessellated or blotched forms. The distinction does not really convey differences in cultivation, although taken as a general characteristic it could be said to divide the warm-growing paphiopedilums from those which will thrive in intermediate to lower temperatures.

It is wrong to evaluate any as cold-growing, although at times this is advocated by those who may have never grown the genus but simply advise from the sales angle. Possibly the coolest-growing species would be *Paphiopedilum venustum* (Wallich) Pfitzer, considered to be the first of the genus to be flowered when grown in cultivation. It is cool-growing in the sense that it may thrive in daily temperatures, extended for preference through night, of 10 degrees (50°F.). It is of little use trying to equate that with the temperatures pertaining to the habitat in the Himalayan region. We are looking at artificial cultivation, which is very different. *P. venustum* has tessellated foliage.

Another hardy and people-resistant species is *P. insigne* (Wallich) Pfitzer, from much the same regions as *P. venustum*. This plant has proved at times that it may be grown as a windowsill orchid, though like most orchids it will not grow for as long in such conditions as it would in a glasshouse. Windowsill cultivation almost always shortens the life of an orchid.

Paphiopedilum insigne has an equal in hardiness under what could be termed adverse conditions. It is *P. villosum* (Lindley) Pfitzer. But some care in trying it in these conditions is necessary. As a species it originates in north-eastern India, Burma and Thailand and over that span of territory it varies in tolerance. These three orchids should be the beginning of any species orchid collection embracing even a part of the total genera.

The plants with tessellated foliage have increased in number in the 1980s by introduction of unknown (or perhaps forgotten) species from southern China. These orchids originate in cool climates but are more easily accommodated in intermediate conditions.

The genus will grow in bark mixes or pure bark. Liquid nutrients need careful handling and revision from time to time. Light applications of blood-and-bone or hoof-and-horn fertiliser are safe and effective, but should not be overdone.

Peristeria (Hooker) 1831

The most commonly grown of the species, of which there are some six or more members, is *Peristeria elata* (Hooker). These orchids should be regarded as essentially warm climate plants, Central American in most instances and from tropical lowland locations. Some plants have been recorded from up to 600 metres (about 2000 feet). The Republic of Panama adopted *Peristeria elata* as their national flower. In that country where it was once so plentiful it has become an endangered species and is now rigidly protected.

As a plant it is morphologically stout, the pseudo-bulbs frequently about 15 centimetres tall (about six inches) and stout to the point where they are almost rounded. The foliage is proportionately large, similar to that of calanthes or lycastes but much taller. This foliage, despite its thinness, is very durable and plants grown in correct environments may retain it for some three years.

To quote *Orchids of Panama*, issued by the Missouri Botanic Gardens:

> This handsome species is the famous 'Holy Ghost' or 'Dove Orchid', the national flower of the Republic of Panama. The popular names are suggested by the form of the column, its beaked anther and the lateral winged lobes of the basal claw combining to produce within the waxy white flowers a figure resembling that of a dove. The plants are found at low to medium elevations in shaded situations on the margins of grasslands and on rocky outcrops in forest, having in former times been very plentiful in the Canal Zone area . . .

Naturally such large plants need space in artificial cultivation, particularly under glass or other materials where the temperature is subject to fluctuations and they need protection. They should be given what are termed warm conditions but the humidity should not approach that needed to grow phalaenopsis. Neither should it be dry. In some ways culture would be similar to that given phaius (refer to *Growing Orchids, Book Three*), the difference being the temperatures at which each grows best.

Pots for cultivating peristeria should be large, the potting material well drained and preferably containing adequate fertiliser to keep the plants healthy. Fertilisers such as hoof-and-horn meal are ideal, applied at the rate of about a tablespoonful worked into the surface of the compost as soon as the plant commences to grow. When the flower spike, which is taller than the foliage, shows, it is also a good cultural process to feed the plant weak liquid fertiliser each time it requires watering. If the correct level of fertiliser is applied the spike will elongate and the buds open in succession over a couple of weeks. Shade should be adequate and the buds not allowed to burn, as they are white.

Phalaenopsis (Blume) 1825

One of the most fascinating genera of cultivated orchids, the plants and flowers can be most disappointing unless they are given the humidity and warmth they thrive in.

Perhaps one could conclude that phalaenopsis can be grown only with hybrids which prefer the same conditions as the genus. However, this book is about species and while phalaenopsis can probably be grown in conditions which few other genera will accept, some sources indicate that certain species need more attention than others.

The genus has been known as long as any other: its history stretches back to the eighteenth century and the innovator of systematic taxonomy and botany. Linnaeus was not the first to know and describe the plants, being predated in this by another botanist and naturalist. The history of the discovery and appreciation of phalaenopsis has been related in *Growing Orchids, Book Three.*

Several authorities on the genus could be recommended to those needing information on its constitution and the plants comprising the group, particularly in relation to the flowers. The plants are distinctive with their usually large broad leaves issuing from a rhizome and lack of pseudo-bulbs; they have the capacity to re-establish refreshed shoots even when all appears to be dead.

Growers and those accepting the worth and rarity of species orchids are saddened by the death of a plant, but for anyone not prepared to design a mode of living for phalaenopsis this event occurs all too frequently.

The genus is confined generally to the islands of the north-west Pacific Ocean, particularly Borneo, Java–Sumatra and the Philippines. Several collectors worked in this area earlier known by the general title of the East Indies. For the orchids from this vast and prolifically endowed area the first designs for artificial cultivation of orchids were formulated; later attempts were made to grow orchids from other regions, needing different treatment, under these same conditions, with inevitable losses. Different treatment was needed.

Anyone who experiences for some time the conditions prevailing in the warmer islands to the north of Australia will find it easy to observe the different climates orchids thrive in. The first thing noticed is the warmth, particularly in the lowland areas where many phalaenopsis originate. The second and equally important feature is the humidity.

Although such authorities as Helen Valmayor, whose specialty is the orchids of the Philippines, quote members of the genus originating as far west as the Himalayas, Tibet and southern China, it is most unlikely that growers would be able to obtain *Phalaenopsis mastersii* (King and Pantling), from Assam. It is one of about six Indian species; few of them have been noted in cultivation, some never.

It is to the species of the Philippines, Sumatra, Java and Borneo that most cultivators turn and indeed most hybridists seek their variations from plants pandemic there. There are some principal members of the genus in this area and these are most likely to be grown.

The genus as a whole may be grown in pots or baskets, or on slabs. The natural habit of the plants is semi-pendant to wholly pendant and it is from this premise that one should begin.

The material used to support plants should be coarse, yet held in control sufficiently to give the roots a firm base in which to thrive. These roots prefer to cling and wood will encourage this. Slat baskets are suitable: it is difficult to detach the roots when repotting is necessary, but repotting phalaenopsis is a radical move generally requiring severing of the roots. The damage should be minimised and repotting attempted only when the roots are in a growing state, preferably from the main rhizome.

The humidity recommended is a constant 80 per cent or more. This is easy to maintain, but it must always be accompanied by a minimum temperature of about 18 to 20 degrees (65–68°F.) or more. Lapses are unavoidable at times, but they should be as rare as possible.

It should be obvious that few genera will tolerate these conditions; if it is intended to ignore the restrictions and grow other orchids with phalaenopsis, there is a price to pay and good results are seldom achieved. Some compromise is worth while, but to whatever degree, it will be reflected in the general growth and appearance of the plants.

Foliage of phalaenopsis is doubly important because it takes the place of the pseudo-bulb system, and it must be as large and healthy as possible to reflect culture in the quality and number of flowers. For example, *Phalaenopsis schilleriana* (Reichenbach f.), a beautiful pink-flowered species and some would say the best of the genus, is capable of producing branched racemes, each carrying up to sixty or seventy flowers on well-grown specimens.

When cultivated in a mixed collection of species orchids, plants may produce only three to ten flowers.

There is something to be said, however, for selecting those species which have few flowers and concentrating on them. The spirit of 'there is always next year' dies hard in all orchid growers.

In the mainstream of the genus appear three or four outstanding species and their worth is never diminished provided they are grown thoughtfully and provided with suitable conditions.

Having noted phalaenopsis growing in almost every region where they are possibilities I have been able to make comparisons. These led to the belief that orchids should be chosen on the basis of natural environments and not made to fit into wanted patterns.

The high humidity, if natural, is not always a guarantee of success. As well as the restrictions on lower ranges of temperature, there is also apparent a restriction on prolonged high temperature accompanied by high humidity.

This may indicate the need for differences in day and night temperatures. It is not easy to lay down figures for this, but it seems to be a necessity. It applies also to other genera, sometimes to a greater degree, and should be thought of as applying equally to all genera which originate in warm climates.

In nature the phalaenopsis occupy situations suited to their habits — the foliage should be allowed to droop, and it should be held on plants for as long as possible, usually culminating in well-grown plants holding up to five leaves. The genus prefers subdued light, with shade of 75 per cent or more, particularly in the warmest point of sunlight. Try to prevent leaf burn, because the persistence of leaves for some three or more years means that any disfigurement is permanent.

Marking of foliage is not confined to leaf burn. In areas where plants are subject to still, moist and perhaps stale atmosphere the spores of fungi proliferate on the surface of the leaves and particularly in the leaf axils of horizontally grown plants. While sprays can be used, it is better to use fans, particularly for flowering plants, to circulate a drying stream of air to reduce surface moisture remaining on leaves. Permanent circulation is essential.

Over-fertilising is rarely a problem with quick-growing orchids like phalaenopsis. Liquid fertiliser should be applied regularly, preferably in weak concentrations, to the root system. Potting material should be minimal and coarse.

Phragmipedium (Rolfe) 1896

A genus principally occurring in South America, a member of the family group cypripediloideae, separated from the genus cypripedium for obvious reasons late in the middle years of orchid culture. Rolfe, the author of the genus, was one of the pre-eminent botanists and taxonomists in the history of the cult.

The genus ranges from Mexico to Venezuela, Brazil, Colombia and Peru to Bolivia. In earlier periods it was variously known as selenipedium and phragmipedium. In the late years of the nineteenth century it was a popular glasshouse genus and proved a hardy orchid, although needing intermediate and at times hot-house treatment.

There are more than twenty recognised species, a few of which grow in Central America.

Garay's revision of the genus during the 1970s changed little. The distinctions which separated the sections of the genus remained. In total it is a warm-growing group, although not tropical. In such conditions an effort would need to be made to ameliorate some of the hotter days from 30 degrees (85°F.) down to a realistic 25 degrees (76°F.). In the best situations that would be an annual maximum, with falls to 15 degrees (60°F.) few and of short duration.

The foliage is similar to that of green-leafed paphiopedilums, and if anything is slightly more resistant to light. Shade of between 60 and 75 per cent may prove sufficient in most situations, but in hot bright conditions this may need increasing to 75 or 80 per cent. It is interesting to note here that two layers of 50 per cent shade, one green, the other black, are usually sufficient to break even the hottest sunlight, but it is wrong to add the two together to make 100 per cent.

Pots and potting material may be similar to that for cypripediums. I have always preferred clay pots to plastic for any genus, but especially for these genera. There is good reason for manufacturers to increase the range of clay pots and for growers generally to return to their use.

As with paphiopedilums, the severing of plants into single growths is entirely wrong; unfortunately it is usually only in this fashion possible to buy phragmipediums. Until the plants have matured into three or four growths it is unlikely they will flower. Those on single growths should be removed.

Fertilisers will need careful selection. Phragmipedium species may not take kindly to some liquid fertilisers and perhaps blends may need working on. Solids such as light applications of blood-and-bone or hoof-and-horn fertilisers may prove best.

Phreatia (Lindley) 1830

This genus of more than 250 species is possibly least known and cultivated of any in the catalogue. Like oberonias, it is undistinguished in its flowering, is principally sub-tropical to tropical and part Australasian in character, extending into Asian countries. Some may regard the genus as of only botanical interest, but possession of any of the species in a collection would enhance it.

As a genus the orchid extends from New Caledonia to northern Australia; the latter has two known species and possibly more. Niugini is rich in the species, and they are found in all the islands of that region including the Philippines, Malaysia and northern India. Some are miniature plants.

One of few references to the genus in the late years of the nineteenth century concerned *Phreatia pusilla* (Bailey), the correct name of which is *P. baileyana* (Schlechter), which grows on Mount Bellenden-Ker at more than 1200 metres (4000 feet). For those who have visited this area of North Queensland the mountain and its neighbouring peak Bartle Frere represent memorable sights.

During World War II In Niugini the author had the pleasure of seeing and collecting this orchid, usually found growing on branches of trees overhanging the rivers and somewhat hard to get. Until identified the plant was a mystery and its plentiful occurrence tantalising. The plant illustrated on page 183 is one of the survivors of that period.

There is a large number of Niugini species, many listed by Schlechter, and perhaps many more which have never been described or named. Most grow at altitudes of about 500 to 1000 metres (1650 to 3250 feet), usually listed as near or on the banks of rivers, principally in light forest rather than the denser rainforests.

The species usually grow in medium light and in cultivation the light could be related to the foliage, which is pale green and with little substance. Temperatures need to be intermediate to warm, preferably about 18 to 21 degrees (65–70°F.) minimum for much of the growing cycle, although in the author's memory night temperatures at rare times could be cool.

The genus, which has little or no pseudo-bulb system, should preferably be mounted on a slab with a backing of fibre to hold the fine root system. Proliferation is by side shoots from the main plant, and when in growth or flowering the plant should not be allowed to go dry. The flower spike is liable to decay if watered once the head shows from leaf axils. Fertilisers should be liquid, sparingly applied.

Pleione (David Don) 1825

This genus is the most cold-tolerant of all orchids. It has been known and grown for most of the history of the cult, but is completely unsuitable in intermediate or warm climates. It almost resembles a perennial plant in its growth habit because it is completely deciduous and renews its pseudo-bulbs each year.

Many species have other peculiar growth phases, some producing adventitious shoots on the upper surface of the pseudo-bulbs; these detach themselves and grow from the corms into mature pseudo-bulbs.

Plants mostly produce single flowers, which develop in the leaf whorl during the primary growth phase in spring each year. Once the flower matures and opens it lasts only a few days and the pseudo-bulb matures over the summer; the plant casts its leaves in nearly all instances, losing the root system at the same time and entering dormancy over the winter months. Occasionally the leaves are persistent, but the pseudo-bulbs flower only once. The old pseudo-bulb may produce more than one shoot from the lower surface, and not all shoots flower. This sometimes gives the impression that the flower stem also grows from the side. In these instances both shoot and flower stem indicate new growth. The species like shade and coolness throughout the cycle.

The genus extends from northern India through to southern China, Thailand and down into Taiwan or Formosa as it is sometimes known. About twenty species make up the genus, but it is possible that total is incomplete.

Potting containers are usually shallow, as the root system is never deep. The plants are naturally epiphytic or lithophytic and very occasionally terrestrial. These shallow containers should have adequate drainage holes bored through the base if they do not have them when manufactured. A very light layer of gravel or coarse flakes of crock material form a good base which will retain some moisture in the growing period, as the containers are apt to dry out too quickly and completely to suit the genus. The potting material recommended is small bark, gravel and shredded sphagnum moss in equal parts. This may need modification.

Once the plant has flowered and the pseudo-bulbs have matured, the pans should be gradually dried out but not quite dehydrated. The bare pseudo-bulbs are dormant until the cycle recommences with the flowering shoots.

Once growth and flowering have started, the plants will take weak fertilisers almost every day and the shallow dishes or pots should not be allowed to dry out. Watering at the wrong time will not suit the genus and dry containers are equally bad in the growing season.

Pleurothallis (Robert Brown) 1815

This very large and diverse genus has incorporated several other genera and species associated with it. It would not be too much to say that it has the most peculiar mixture of plants possible to imagine. Many flower from the axil of the leaf supported on a stem, others flower from the rhizomes. They are epiphytic, lithophytic and occasionally almost terrestrial or approaching saprophytic.

The general habitats are Central and South American, the plants growing in moist areas either on lowlands or in cloud forested mountains. Many species could be considered of only 'botanical' interest to those growers who prefer large plants, and flowers in proportion. But for the species fancier they represent a large section of desirable orchids, hard to get and sometimes not very reliably named.

In the history of the genus the species have been known by the generic titles of acianthera, acronia, anathallis, brenesia, centranthera, crocodilanthera, duboisia, dubois-reymondia, humboldtia, kraenzlinella, myoxanthus, otopetalum, restrepia, rhyncopera, specklinia and talpinaria. All this taxonomic change has been incompletely publicised and many plants such as *Restrepia dayana* should be known by another name. It had this name when bought by the author and was subsequently identified as *Restrepia elegans* (Humboldt, Bonpland and Kunth) but has since been transferred to the pleurothallis, where it apparently is known as *Pleurothallis ospinae* (R. E. Schultes).

Pleurothallis should be grown as ordinary epiphytes in bark mixes. The only restriction is size of containers, which should be kept as small as possible considering the size of the plants. They are not common orchids in average collections and distribution of species is infrequent. Small pieces should not be used for propagation because they take too long to re-establish as flowering plants, although sometimes single detached leaves will grow in moss.

Lists of species are available from nurseries, but ordering plants from these is not the best way to buy. If possible the plants should be bought as growing propagations, sometimes available at orchid club meetings or shows. Most species prefer intermediate to warm conditions.

There is frequently no real lead as to the needs of plants, but the sure way to do the right thing is to watch for new growth rather than trust to chance watering. As a genus the plants are extremely hardy, and more damage can be caused by over-watering than under-watering.

Flowering period, important as it is, may not be known or is given wrongly, but with most it will be late summer to autumn and the plants should be watered and fertilised with this in view. Most favour bright light, and 60 to 75 per cent shade is adequate.

Polystachya (Hooker) 1825

An infrequently grown yet almost universal genus, absent only from Europe, Russia and China. While many species are of African origin, the remainder are spread thinly through other sub-continents and continents in both hemispheres. Again it is possibly a 'botanical' genus, with some species really worth more attention than they have been given in the past.

One hundred years ago the genus as a whole was described as comprising 'novelties' as distinct from the cultivated genera of the time, which were usually selected for their beauty rather than unusual characteristics. Most of the known species grown came from Africa. E. F. Hennessy illustrates Joyce Stewart's *Orchids of Africa* with several beautiful reproductions of African polystachya.

The genus as a whole is epiphytic, occasionally lithophytic, and most species exhibit the pseudo-bulbs and several leaves one would expect of that classification. There is, however, a great difference in flower presentation, ranging from erect spikes with equally spaced flowers to those bearing umbels and straggling pendant racemes.

Twelve species grow half a world away in Brazil, but little information about them is available. One species grows in the Philippines. Its principal claim to fame is possibly in having fifteen synonyms. It is one of few which grow in the Asian sphere. Possibly the first of the genus flowered in cultivation was *P. luteola* (Swartz) Hooker, which produced its blooms for Loddiges in 1820. It was a Mexican orchid common also in other countries. The pseudobulbs apparently flowered several times in similar fashion to some dendrobiums.

Nearly all the species are considered warm-growing, some intermediate, but none are recommended for cool growing in temperatures below 10 degrees (50°F.). They need the usual light humidity associated with epiphytic orchids, particularly when in their growth stages. The plants flower on mature growth, which in the normal pattern would occur in summer to autumn in temperate climates. Sub-tropical and tropical climates may prove different, and appreciation of this factor is necessary when cultivating such a widespread genus.

Light plays an important part in growing epiphytic orchids and few polystachya will tolerate the dark conditions which are usually provided for paphiopedilums. In such environments the plants should be suspended rather than benched.

Pots may be either plastic or clay. Fertilisers are necessary in the growing stage, but should not be heavily applied.

Promenaea (Lindley) 1843

These orchids appear to be solely Brazilian, the number of species varying considerably according to which authority is consulted. But there are more than twelve, and perhaps as many as fifteen. Hoehne in his survey of the genus in the 1950s listed sixteen. The genus is widespread in the southern states of Brazil, most growing in a climate engendered by the soft moistness of the winds from the Atlantic falling on the mountains of the coastal section, with a little extension into the hinterland.

The greatest aggregation is in the Santa Catherina region in southern Brazil, in the warmth of latitude 25 degrees south. While most of the species are considered cool-growing they do not thrive in temperatures below 10 degrees (50°F.) and are slow to grow and flower in it. They are not warm-growing either, and will quickly lose vitality in areas with warm summers beyond temperate conditions.

The best-known of the genus is *Promenaea xanthina* (Lindley), which was originally sent to England by Gardner from the Rio de Janeiro region about 1837. Lindley named the introduction *Maxillaria xanthina* and later formed the generic title promenaea for the flower which was ill-suited by the name maxillaria. Discoverer of this species probably was Descourtilz, an early collector of plants in Brazil, including *Sophronitis coccinea* (Lindley).

Other species seldom feature in collections. Some are a little difficult to acclimatise, but most failures are attributable to lack of understanding and unsuitable environments.

Although most of the genus is epiphytic and seldom found in other situations, it is not moist-growing like masdevallias. It would be better suited by conditions created for *Sophronitis coccinea* and its kind. Those who have cultivated this orchid will know how to treat it to get results, principally by not overwatering the plants even when they are in the growth stage. In this stage they should be given only enough moisture to stop the older pseudo-bulbs from shrivelling beyond the slightest of ribbing.

In growing *Promenaea stapelioides* (Link and Otto) Lindley there was some difficulty until I adopted the double-pot system and infrequently watered the plant, being careful not to water the foliage at any time. The change was rewarding. The same system also worked with *P. xanthina*. Much may depend on the humidity of the enclosure.

Most of the genus flower in summer, occasionally at other times. The flowers which appear out of season frequently damp off and the best thing to do is to treat the plant as unflowering and let the buds decay.

Light requirements are the same as for cymbidiums and other hardy orchids, but the plants should not be hung near the glass. Fertilisers applied sparingly do more good when flower buds are noted. But keep the buds and leaves dry.

Renanthera (Loureiro) 1790

A genus with habitats extending from India to islands in the Niugini region. It is principally warm-growing, and includes tropical species. The Indian species *Renanthera imschootiana* (Rolfe), growing at 1000 to 1500 metres (3250 to 5000 feet) almost belongs to the cool-growing strata of plants. They are vandaceous orchids.

The plants are not large. Four of the twelve or more species grow in the Philippine islands and most are short. These, with the Javanese and Sumatran members, belong to the tropical section, where most of the genus is concentrated. Most grow at low elevations and in bright light. The flowers are principally red, sometimes buff-yellow overlain with red.

They are usually spectacular bright orchids in their own environment, showing out starkly in the bright sunshine of the tropics. A short synopsis of the history of the genus and its association with growers in the nineteenth century is given in *Growing Orchids, Book Three*. They are monopodials which normally grow in bright conditions and frequently on the bare trunks and branches in what are considered arid situations, subsisting on rain and moisture carried in the air. The substance of the plant is sparse.

The plants may produce as many as three or four racemes of flowers from the internodes of the stems and on opposite sides from the leaf axils. They are nonsupportive and must normally be tied or staked erect in cultivation. If grown from low beginnings among other plants they may not flower until they have cleared their surroundings and reached bright light, occasionally almost at the roof of enclosures. They are suited to growing in shade-cloth protection of about 50 per cent. Under this canopy the leaves may burn if soft, but it is conducive to free flowering.

In cool conditions the leaves may develop black spotting such as may be noted on any genera in unsuitable temperatures. Except for species from cool climates or high altitudes, the genus prefers warm conditions, but even these may at times be found unsuitable if there are long periods of dull weather or coolness.

The root system is almost entirely aerial, so watering may be minimised by high humidity or by encouraging longer roots to enter into potting material. The usual tendency is for the roots to remain aerial and they may have to be trained to do what the grower wants. The type of container is immaterial, but the potting medium should be coarse. If in the growing period a little moss topping is saturated with fertiliser each time the plant is watered the roots may look for this.

Restrepia (Humboldt, Bonpland and Kunth) 1815

It is somewhat difficult to define where this genus begins and pleurothallis ends. Some species previously known as restrepias have been removed to pleurothallis and others remain in the restrepias. This is despite the obvious resemblance of one to the other.

It is apparent that many of the species listed and labelled as restrepia should be transferred to pleurothallis. The work load of taxonomists grows each year, but at times it seems there is more concentration on species and genera in the hybridising section of the craft and less on the botanical interest group.

The restrepias are principally of a form which is distinctive from the pleurothallis by flowering from the petiole below the leaf-like pseudo-bulbs. If there is any remarkable distinction it could be based on this as well as floral characteristics.

Restrepias which are commonly grown are few in number. In the collections of the earlier years of the twentieth century, when the emphasis was on larger flowers and brighter colours, the whole of the largest section or genus of orchidaceae was passed by. The restrepia shown on page 187 is perhaps known also by the generic name pleurothallis, flowering from the petiole like *R. antennifera*.

In cultivation the genus restrepia or pleurothallis appears to be one of the easiest of the intermediate to cool-growing orchids to maintain in good condition. Fine potting material suits the rather fine root system. It should be well-drained: the dampness preferred by many of the South and Central American genera can be almost forgotten.

New leaves are formed on some species throughout the year, perhaps at a slower rate in the cooler months, but growth is constant. Flowering is usually in early summer, but some species will flower more than once in the annual cycle.

Shade is not so much a worry with broad-leafed forms such as the plant illustrated and they will tolerate bright conditions in cooler climates down to the 50 per cent level. In order to get the correct environment the plants should preferably be suspended in the brightest part of the glasshouse or enclosure.

In most collections the suspension or hanging of large numbers of plants presents problems. Correct cultivation should outweigh possessive desire at times. Do not suspend plants from roofs not built to take this sort of load.

Restrepia leaves detached from the rhizome will produce new plants from the petioles if placed in sphagnum moss with the blades protruding.

Rhyncolaelia (Schlechter) 1918

These orchids, comprising only two species, are confined to the Central American region, sparsely in some regions and growing relatively freely in others. In the late years of the twentieth century they have become rather rare and unless some measures are taken in the countries of origin they will perhaps become extinct.

Both plants have similar morphology, with a rather hard-looking, coarse leaf, seldom more than single, topping a slim and hard-looking pseudo-bulb about 16 to 20 centimetres (6 to 8 inches) tall. The habit of growth of both is close set. The plants are economical on space and with a greyish appearance.

These orchids are more suited to pots than to slab mounting or other systems. The potting material should be coarse and capable of dehydration to a degree in conformity with conditions of the natural habitat.

Rhyncolaelia digbyana (Lindley) Schlechter will probably always remain for many growers *Brassavola digbyana*. Under this name it was used to influence modern orchid culture to a degree that cannot be overstated. The flower is distinctive because of the fringing of the labellum and occasionally the petals also in a recognised variety.

Rhyncolaelia glauca (Lindley) Schlechter also went under the brassavola label in the same manner until the alignment with that genus was recognised as impossible and both were transferred. It is a plain relative of the flamboyant *R. digbyana* and has never been cultivated as freely as its flower merited. If anything it appears easier to grow than *R. digbyana*. Both flower in late summer in some climates and spring to summer in others. Both species are scented.

Neither species, despite their warm origins, appears suitable for tropical cultivation; they succeed in warmed glasshouses or sub-tropical to lightly tropical climates. Growth must be completed to maturity by late spring or summer, and it is hard to get the plant to follow this pattern in warmer zones. The tendency there is for climatic influence to force the opposite system.

Two points about the flowers of *B. digbyana* are worth mentioning. Seldom, if ever, will a grower flower a plant with the labellum correctly oriented to the rest of the flower. It appears always to have a twist. The second is that if the flower is fertilised and the capsule matures it will have a long 'beak' or 'neck' at the end of the capsule.

Culture of both species is as for cattleyas. They have the same periodic dormancy and sheath from which the single flower emerges. Fertilisers are necessary only when the plants are in growth, and should be lightly applied to finish texture and body in the flowers. The genus was more fully covered in *Growing Orchids, Book Two*.

Rhyncostylis (Blume) 1825
Saccolabium (Blume) 1825

These two genera are allied to sarcochilus and other vandaceous orchids, as can be seen when their foliage is compared. Like the aerides and ascocentrums, the plants are robust epiphytes from the Indo-Asian area. In common with vandaceous orchids generally, once established, plants are intolerant of disturbance.

Flowering habits of all these medium-sized orchids are similar, with some rhyncostylis and aerides producing dense racemes of flower, usually from within the leaf axil. For example, *Rhyncostylis retusa* (Linnaeus) Blume was given the epithet or colloquial name of 'Foxtail Orchid'. In British terms it adequately described the appearance of the bright pink-purple flowering mass.

Most of these orchids depend on environment for adequate rates of growth and flowering capacity. They cannot be hastened by fertiliser alone. They belong to the intermediate climate range, preferring temperatures not lower than 15 degrees (60°F.) but tolerating occasional lapses to 12 degrees (52°F.).

Shading should be in the medium range of about 50 to 70 per cent and there is little need to worry about bright conditions. Although the foliage may become slightly lighter in colour, the tolerance is high for anything other than direct sunlight. A number of our cultivated genera tolerate this, so a moderately mixed collection can be grown in these bright conditions.

Pots usually are unsuitable, as root systems should be encouraged to become at least partly aerial. Generally these roots will be short and quickly cease growing if temperatures drop and remain low for any length of time. Containers other than pots include baskets of either wire or wooden slats, lined with a layer of coarse shade cloth, or husk of stag or elk horn fern. Potting material should be coarse and suited to the climate. If bark develops growths of fungus or other deleterious things which inhibit root growth, a substitute like large chunks of charcoal should be tried.

Rhyncostylis retusa (Linnaeus) Blume, *R. gigantea* (Lindley) Ridley or *R. coelestis* (Reichenbach f.) are the species usually obtainable, but most saccolabiums have been transferred to other genera.

Fertilisers for actively growing plants may be solid or liquid. If using solids, make a small bag from discarded nylon hosiery, put the fertiliser in the bag and water through it. Remove the bag once the plant stops growing or flowering.

Robiquetia (Gaudichaud) 1826

This genus of vanda-like plants is not large, having fewer than twenty members which are spread over the Asian region and down into Niugini and northern Australia. Many are transferees from other genera, but the general appearance is that of rhyncostylis and other small vandaceous monopodials.

The species are mostly tropical orchids, occupying epiphytic positions north and south of the equator, but confined into a rather narrow band of habitats occurring from about 200 to 1000 metres (750 to 3250 feet).

While the flowers in their non-opening habits are not as attractive as aerides and ascocentrums, the dense flowering pendant spikes are usually brightly coloured and popular with growers.

Two Australian species, *Robiquetia tierneyana* (Dockrill) and *R. wassellii* (Dockrill), from North Queensland, are probably least attractive because their flowers are small and dull coloured. Niugini has several species and the Philip pine islands five. Few if any species occur in countries other than that of their origin. *R. tierneyana* has been renamed *R. gracilipes*.

Species growers are not deterred by the fact that the flowers are small and not as bright as the aerides, ascocentrums and vandas. Some are concerned with numbers in their collection while others have variety as their ambition.

Growing these small vandaceous plants is straightforward because they will grow in pots, on slabs of treefern or in wooden baskets which may be hung against a wall. The plants will move naturally toward light so this should be considered and plants not left in shaded positions. Most are bright-light orchids and will tolerate up to 50 per cent shade. If grown too dark the plants may produce shorter spikes or none at all.

Hanging plants, of course, are always a problem in enclosures because they take up so much space; it is possible that if the plants are benched they may be too frequently watered and not get enough light. The fact that robiquetias can hang in any position makes them ideal subjects for outdoor or almost totally outdoor conditions.

Potting material should be coarse and the scope given for much of the root system to be aerial. The whole of the vandaceous section tend to produce aerial roots.

When the roots are allowed to wander about, the matter of fertilising becomes difficult and the root systems should be encouraged to partly inhabit containers, preferably their own. This is done by training them gently in the direction desired. Fertilisers may be solid if some of the root is contained; otherwise the usual soluble fertilisers applied in the growing season are sufficient.

Rodrigueziella (O. Kuntze) 1891

A small genus confined almost solely to the southern regions of Brazil. The species had a chequered career through the list of names and the genus may be referred to in outdated literature as theodorea. The species *R. gomezoides* had many synonyms and superseded titles, among which were *Theodorea gomezoides* (Barbosa-Rodrigues), *Gomesa theodorea* (Barbosa-Rodrigues), *Gomesa theodorea* (Cogniaux), *Hellerorchis gomezoides* (Barbosa-Rodrigues), *Theodorea paniculata* (Brade), and *Hellerorchis paniculata* (Brade).

The plant would not be considered worth a place in collections of some growers, yet it is highly prized in others. The genus is not widely cultivated and the species reviewed below is the most widespread. It grows in six Brazilian states or subdivisions, all of them in southern parts of the country.

The climates of origin of all species are similar, principally in lower elevations of the foothill country of mountain systems. They could be regarded as cool-growing, yet probably all would benefit from temperatures above 12 degrees (52–54°F.). The term cool-growing may be misleading, and cultivation should be based on appreciation of individuals and not on the thermometer.

The rodrigueziellas are epiphytic, related to a fairly large group of Brazilian orchids in what could be called the hard-growing core. All orchids in this category are summer-growing types that should be given fairly dry dormant periods over their non-growing season. The total system should be reversed once plants enter their growing phase when watering should be frequent. The foliage is tough and the habit similar to that of oncidiums, with slightly longer sections between pseudo-bulbs. This habit is taken to extremes in the related species *Rodriguezia decora* (Lindley) Reichenbach f., which may have sections of stem or rhizome between pseudo-bulbs as much as 10 to 15 centimetres (5 to 6 inches) long.

Potting material for all members of the genus is bark-based, a judicious mixture of fine and coarse so that drainage is perfect. The root system must be preserved over the dormant period and the signal to increase moisture slowly is given by the appearance of new growth.

Shade for the genus should be of 50 to 75 per cent, preferring the lower density unless there is any indication of leaf-scorch or yellowing. The plants may be benched provided they are not overhung by foliage of other plants. Pots may be either clay or plastic with the same potting material. Treatment should be adjusted to hanging and benched specimens.

Rossioglossum (Schlechter) Garay and Kennedy 1976

This small genus is based on the most recognisable of the species once known as odontoglossums. It is Central American in origin, spreading through more than one of the states of the region and with only a few species. It thrives in cymbidium mixes.

Best known of these is *Rossioglossum grande* (Lindley) Garay and Kennedy. At times other species, remarkably similar, are mistaken for this orchid and confusion has been the order of the day. This was dispelled many years ago by the English grower and authority Peter Black. Despite this, the mistakes still occur.

Probably one of the easiest orchids to grow if sufficient care is taken to accommodate it, *R. grande* is a temperate zone plant which slots very well into species selected for that characteristic. In conditions that are too warm the plant sulks, and if tried in cold conditions it also fails. Too frequently it is mistaken for a plant that grows cold. It fails as a house plant, although sometimes recommended.

There is much difference in the colour pattern of the petals and sepals in each species and without definite identification or knowledge the labels on plants should not be changed. *R. grande* itself is variable.

The growing period is from late spring and this is where an early start shows benefits. The plant may not proceed very quickly to flowering, but the early start ensures steady growth to production of flower spikes on the new growths.

In James Bateman's *Orchids of Mexico and Guatemala* the flower spike is shown as appearing to come from an older pseudo-bulb, whereas in fact it comes up within the enclosing leaf bracts of the new growth, often apparent about the end of January on the earlier flowering species *R. insleayi* (Barker) Lindley. Some doubt still lingers with authorities about *R. williamsianum* (Reichenbach f.) Garay and Kennedy.

Once the plants of any of the species have grown and flowered and have obviously completed their make-up they should be put where they get only an occasional splash or no water at all until they show new shoots from the bases of mature pseudo-bulbs. Even then there should be no haste to water the plants; after a week or two they should be slowly moistened and brought back to growing stage. If watered during dormancy the root system may be damaged throughout and a new one must necessarily be produced.

In active growth the plants should be sparingly watered and regularly fertilised with weak solutions of nutrients. Once watered they should be left until only moist, and then again watered heavily. It is best to apply the liquid fertiliser a couple of hours after a thorough watering. The secret of success is correct temperature and complete dormancy. The shade factor is about 60 to 75 per cent and humidity 50 to 60 per cent.

Scuticaria (Lindley) 1843

This small and unusual genus is limited in its habitats to Brazil and Venezuela and supposed to also grow in Guyana. The species are few and epiphytic and seldom noted in collections. The original importation into Britain was of *Scuticaria steelei* (Hooker) Lindley in the 1830 decade, when it was known as a maxillaria.

Those plants came from Demarara River in Guyana. Little is known about this area in the late years of the twentieth century and whether the orchid is still found there. Brazil and Venezuela are regarded as the source of plants when they are available, with Brazil having the greater number of species.

The plants themselves are morphologically unique. One has only rudimentary pseudo-bulbs topped by long cylindrical and tapering leaves which have been known to grow to 1 metre long (3 feet). The flowers, generally in singles or occasionally two or three, are borne from what in reality is a rhizome system and grow from between the pseudo-bulbs, which are separated by short sections.

The climate in which plants with such foliage usually occur is bright and rather dry, but the scuticarias grow principally in light forest, sometimes in quite shady situations. They are usually found in high rainfall regions on the lower slopes of foothill country rather than at high altitudes.

Hoehne had perhaps the best experience of the botanists and illustrated three species in his *Flora Brasilia Orchidaceas. Scuticaria strictifolia* (Hoehne) broke the usual pattern of the genus by growing erect-leafed short pseudo-bulbs and having a marked cochleate labellum. In his illustrations of *S. hadwenii* (Lindley) Hooker and *S. steelei* (Hooker) Lindley, the marked differences in morphology are apparent. *S. hadwenii* has little or no pseudo-bulbs, while *S. steelei* has definite pseudo-bulbs which shed the sheathing bracts as the leaves lengthen. Hoehne also showed the leaves of *S. steelei* curled about the page until he ran out of space.

In cultivation all these orchids should be grown suspended, preferably on slabs of treefern or other mounts, so that the foliage can hang free. The exception may be the erect *S. strictifolia*, which would present an alternative. All should be grown in bright light, 50 to 60 per cent.

The climate for all the species is intermediate to warm, and preferably not below 18 degrees (65°F.). The plants grow over the summer months in cooler climates and may be unreliable in tropical conditions. The flowers appear in summer to autumn.

Fertilisers to feed the medium-coarse root system should be applied over the growth period and until flowering, when the plants will become dormant. Resume active cultivation once new growth commences.

Sigmatostalix (Reichenbach f.) 1852

A Central and South American genus numbering about twelve species and seldom cultivated. An asset to species orchid collections which aim to include uncommon genera. Some time could be spent arguing the merits of such collections, but, like philately, discussion would be pointless. Each individual has preferences.

Some of the genus have the facility of producing racemes which, in fashion similar to some masdevallias and other genera, flower again and again from the same stem.

The genus is spread, with members not all growing in the same region. Some have an appearance similar to cycnoches flowers in the arcuate column, although in most instances the flowers are much smaller than could be imagined from the comparison.

As naturally occurring orchids the species are sparsely represented in the flora of most countries where they grow. In this sense, the species *Sigmatostalix guatemalensis* (Schlechter), with its oncidium-like morphology, has been noted quite rarely in that country. It grows at 1000 to 1700 metres (3250 to 5500 feet) in humid and wet rainforest. It could be deduced from this that it is in little danger from land clearance. But, if rare, it is possible that destruction of forested areas by any means could drastically affect its survival.

Only two species occur in Brazil, both in Amazonas, in the northern section of this vast country. Little information is available except that they both grow in a lowland climate which is very hot and humid. This at least gives a lead to their culture.

In spite of the restrictions on the export of orchids from Brazil, there seems to be an endless stream of plants under the permit system. It is only by inclusion of such orchids as sigmatostalix with mixed collections of plants that they will be available from time to time. Amazonas being what it is, huge and thickly covered and with intersections of water wherever one turns, whatever plants of the genus occur are well protected.

If they are cultivated, temperatures of not less than 15 to 18 degrees (60–65°F.) should be planned, with the upper limit quite flexible. The resemblance to oncidiums should be a guide to potting methods and materials. The root system is not coarse, but more like that of the finer to medium oncidium root systems and probably dense. Only infrequently have I noted plants in cultivation. Shade cloth of 50 to 60 per cent should be sufficient.

Watering and fertilising should follow the pattern of growth and dormancy. Flowering season is in summer to autumn and once past that the plants revert to dormancy.

Sobralia (Ruiz and Pavon) 1794

Central and South American, the sobralias are quite different from orchids common to much of the area where they grow. The genus numbers more than twenty, although few of them are grown. The species more commonly noted are the large-flowered types which produce their blooms over most of the warmer months of the year when flowers are not noted on epiphytic orchids. The sobralias are principally terrestrial, very occasionally growing above ground on decaying forest material or in rock formations containing similar concentrations.

The genus as a whole has been much neglected throughout the entire history since its introduction in the early 1830s. *S. decora* (Bateman) appears to have been the first to reach growers in England and Europe; it is pale pink with a rose-coloured labellum.

It is difficult to account for the present neglect of the genus; it is possibly because the plants, which are similar morphologically to arundinas, with tall, thin, bamboo-like growths and no pseudo-bulbs, travel poorly. One importation which I saw in the 1980s was a total write-off after quarantine procedures and isolation.

The root system of the genus generally is coarse, like that of phaius. The potting material should be modelled on that genus, with a base mix rich in humus, free draining and not soggy after watering.

Flowers of the metre-high stems vary from rose-pink to white and yellow cattleya-like blooms produced in succession on the head of the stems to the number of four to six on well-grown plants. They like moderate protection such as 75 per cent shade cloth, and temperatures preferably no lower than 15 degrees (60°F.) through the colder months. Pests such as red spider seem to favor the genus and steps should be taken to counter them.

Darwin, in his work on the fertilisation of orchids, had this to say of a pollination:

> Mr Cavendish Browne informs me that he saw a large humble-bee enter a flower of Sobralia *macrantha* in his hot house and when it crawled out it had the two large pollen-masses firmly fixed to its back, nearer the tail than the head. The bee then looked about, and, seeing no other flower, re-entered the same one of the sobralia, but quickly retreated, leaving the pollen masses on the stigma, with the viscid disc alone adhering to its back. The nectar of this Guatemalan orchid seems too powerful for our British bee, for it stretched out its legs and lay about for a time as if dead on the labellum, but afterwards recovered.

Sobralias need large containers to accommodate a prolific root system and appear well suited to sub-tropical climates, where they could be used as garden plants.

Sophronitella (Schlechter) 1925

A small orchid, isolated as sole representative of a genus, it has been for long a favourite with many growers. In the early years of its history it was classified as a sophronitis and was known as such through to its renaming by Schlechter. *Sophronitella violacea* (Lindley) Schlechter is native to Brazil and no other country, being found originally in the Rio de Janeiro area.

That a beautiful and easily grown orchid has always been rare in collections is remarkable. It is cool-growing, thriving in low temperatures above 10 degrees (50°F.). It is one of the more attractive winter orchids in temperate climates. Although it takes a number of years to grow into a large plant such as that shown on page 192 every minute of the time becomes well spent when looking at it in flower.

The *Botanical Register* of 1841 was probably the first notice of the plant in cultivation, where it was described as 'a pretty little epiphyte with violet flowers, for which I am indebted to Mrs Cannon, of Stratford Green, in whose hothouse it flowered in the beginning of February'. That month, of course, in Britain, is just past midwinter and that is the period in which it normally flowers.

While known as a sophronitis it lacked attention, possibly because of its size and colour. *Sophronitis coccinea*, also known as *S. grandiflora* by many growers, was larger and brilliant red, against the magenta-violet of *Sophronitella violacea*, with its small flowers and less striking appearance.

There is little resemblance between the two genera other than their size. Both are regarded as almost miniature orchids. The foliage of sophronitis is broad or narrow, depending on which species is considered. That of sophronitella is narrow, and channelled, and the pseudo-bulbs are covered with the white sheathing remnants of the bracts many years after the pseudo-bulbs have flowered.

In cultivation *Sophronitella violacea* is a particularly easy subject. It may be potted in a plain bark mix, fairly fine yet not too much so. It may also be grown open-rooted and tied to a suitable mount, to which the roots, never quite as long or stout as those of sophronitis, will soon cling. The root system is capable of damage by overwatering, whether plants are potted or open-rooted.

When flowering is over, the plant does not enter dormancy for any time at all, but immediately sends out shoots from the base of the pseudo-bulbs, which take almost ten months to mature and flower in their turn.

Shade should be of 60 to 75 per cent, but the plant is unlikely to burn in brighter light. Fertilisers should be applied once growth is noted. Too much, too soon and too strong halts root development.

Sophronitis (Lindley) 1828

A genus essentially Brazilian but possibly also growing in neighbouring states to the east and south of that country. There are a number of species, but how many is a matter of dispute amongst authorities.

Some species are cool-growing while others are suited better by a temperature a few degrees higher. The minimum is preferably above 10 degrees (50°F.), but I have grown and flowered *Sophronitis coccinea* (Lindley) Reichenbach f. outdoors as an epiphyte growing on casuarina trees at temperatures which went as low as 6 degrees (42°F.). It was not at its best at these low temperatures, but it grows and flowers very well in a minimum of 10 degrees (50°F.) in protection.

Over many years I have been at pains to identify the plants I grow. This of course extended to sophronitis. It would appear there are four variations of the red-flowered species known as *S. coccinea* and I am still undecided if they merit separate names. However, they are so regarded.

From the beginning of the calendar year where I live, the first to bloom is *Sophronitis pygmaea*, which is unrecognised taxonomically. Both the plant and the flowers are very small. It opens its flowers in January to late February, midsummer in my southern hemisphere region, and is classed as a degenerate form of the next in line, *S. mantiqueirae*. This flowers in May–June. The next to flower is *S. brevipedunculata*, which is almost a month later. Then comes what is recognised as the true *S. coccinea*, which flowers in September–October. The species known as *S. wittigiana* is different again because of its rose-pink flowers, and they open in August and September. All these species have been subject to investigation by J. A. Fowlie but his taxonomy shows so much variation in colour of flowers, shape and colour of leaves, and habit of growth that it is too confusing for the ordinary orchid grower.

In addition to this, I believe that one has to grow and study this genus over a number of years to come to any conclusions. However, this authority is the only one recent addition to knowledge of the genus and it must be accepted even if his judgement, based on morphology, leaves and flowers as deciding factors, does not bear out the forecasts and expectations under cultivation. I prefer to adopt my own conclusions based on flowering times, but I believe that most should be reduced to varietal level.

Several other species are listed. They include *S. brevipedunculata* (Cogniaux) Fowlie, *S. pterocarpa* (Lindley) and *S. cernua* (Lindley). The first two are variants of the *S. coccinea* form and colour and a rather straggling habit in *S. brevipedunculata* which results in haphazard growth pattern which could be termed a clump. In *S. pterocarpa* the habit is normal, straightforward side-to-side proliferation resulting in a string-like habit that is supposed to be a distinctive feature of some of the *S. coccinea* complex. It also has compressed pseudo-bulbs.

S. cernua is of smaller habit, both plant and flowers, with the flowers appearing in numbers up to seven or eight on a raceme issuing from the head of the pseudo-bulbs, as in all the genus. The flowers are orange-gold rather than yellow or red. It has another related species apparently originating in Bolivia, which is if anything smaller than *S. cernua* but has the distinct habit of that species. It is tentatively named *S. acunae*.

All sophronitis are easily managed in orchid collections. The plants are small, neat in their habit and need no control. They may be potted or grown as open-root epiphytes and I favour the latter method. I use both, but do not use over-large pots.

In cultivation the only part of normal procedures to watch is watering. Growth commences in most instances months before flowering and the plants never at any stage of their annual cycle are moist-growing species. Dryness is the best condition, with a complete saturation about once a week. If flower size is a consideration, a careful fertilising program from the time the buds emerge from the leaves will give results. It should be noted also that the flowers grow to about half as big again under good culture and conditions.

In their habitat the plants grow either on rock surfaces or on trees and either open-mesh benches or suspension will suit them. Air circulation should be considered also, with either interior or introduced outside air.

Some growers use sphagnum moss as a potting medium, but in inexperienced hands it may become too wet. For beginners, plain bark mixes are best with no need to worry about drainage. The bark should be medium fine but not too fine and this should be soaked about once a week. Humidity of about 50 to 75 per cent is required.

Shade should be moderate, preferably about 60 to 75 per cent over warm months and reduced to about 50 per cent in dull periods. They are hardy orchids and worth every minute spent on their care.

Spathoglottis (Blume) 1825

Among the earliest cultivated orchids brought into Britain and Europe from the 'East Indies', this genus of terrestrial species originates in the wide area of India through southern China, the island archipelagos of the western Pacific through to Samoa. There is remarkably little variation over this area, which is principally tropical to sub-tropical, and the species are usually considered orchids for warm climates.

In artificially created environments neither the plants nor flowers realise their true potential, but it does not appear to halt their cultivation as glasshouse orchids. There are more than forty species, but not all are worth growing as a main genus. They add to the flowering potential of a group of mixed species, but never excited the orchid world as did so many other genera. The situation could be different for growers in sub-tropical and tropical regions, where spathoglottis may be grown as border plants in gardens. In such situations they would be too soft to withstand direct sun.

Spathoglottis plants are calanthe-like, but the leaves of most species are more slender and have a longer petiole than a calanthe leaf. They are veined verti cally, sometimes markedly so. These leaves where they are subject to wind or other damage become straggly and untidy and should be regarded as a vital part rather than one which suffers damage without loss of vitality. Under glass, shade cloth of 60 to 75 per cent is advisable.

The plants have pseudo-bulbs and these should be above surface rather than below. In replanting it is important not to bury them because it is from them that the flower stems arise. They may decay before maturity if sodden.

Potting material should be similar to that for phaius or calanthes or any other terrestrial genera. It should be open in texture, yet able to retain sufficient moisture to last between waterings. While spathoglottis have pseudo-bulbs, they do not possess the qualities of epiphytic orchids and cannot sustain the large foliage without moisture.

Spathoglottis flowers from many sources have much the same shape, but colours vary through the species from white to purple, sometimes each species having variations. *S. plicata* (Blume), a common Philippine islands orchid, has both white and beautiful pink forms. It is widespread and extends through much of the range of the genus down to the Solomon Islands.

The number of flowers on the heads of the long stems varies, ranging from three or four up to more than twelve, which open successively as the lower flowers die.

As glasshouse plants they need warmth, preferably not less than 15 to 18 degrees (60–65°F.), withstand heat very well and flower variably from autumn onward.

Spiranthes (L. Richard) 1818

One of the genera with representatives in every country on earth except the African region and a small part of eastern Asia. Botanists and enthusiasts have had a wonderful time sorting out the species and allotting names according to their vision of the genus. It would almost be possible to fill several pages with their synonyms.

Some species are international in distribution, others restricted to certain areas. All or almost all are terrestrial herbs which grow, flower, defoliate, enter dormancy and then recommence growth as the seasons change. In general a recognisable feature is their terrestrial habit and the underground tuber which is the base for growth and flowering. It is impossible to give the variants from the theme by name or to verify their nomenclature.

Some idea of the habitats may be gauged from an entry in *Orchids of Jamaica* by Fawcett and Rendle to the effect that they had been found 'on stone walls, on rocky banks and in peaty soil in woodlands'. In contrast, another entry reads '. . . in shade, in leaf-mould, in woods and amongst mosses on rocky banks'.

There are more than 300 species in the genus but until someone takes a revision in hand much confusion will continue. In cultivation the plants are rare, principally because of the common emphasis on epiphyte culture and concentration on hybrids to the exclusion of species orchids.

The original designation of the plant on page 195, *Spiranthes navarrensis* (Ames), was *Stenorrhyncos navarrensis* (Ames). It is a Panama orchid from cloud forest at about 2000 metres (about 6500 feet). At this altitude one would expect to find orchids transferrable to cool climates, but *S. navarrensis* is better suited by intermediate culture and pot grown, as a terrestrial should be.

It is doubtful if any of the genus would grow epiphytically, although at times plants may be found in such situations in aggregations of leaf and forest detritus.

Potting material for artificial cultivation should comprise some of this material, but it should be of open texture giving free flow to excess water. In the matter of shade, some 60 per cent upward should be tried and plants watched to guard against burning. Fertilisers should be given sparingly in the growing period and ceased altogether when the flower spike has died off.

Stanhopea (Frost ex Hooker) 1829

One of the most extraordinary genera in the orchid floras. It has no distinctive plant features yet is distinctive in variations of flower formation. To look at the flowers is to marvel 'why and wherefore'. The native races of the Americas were equally intrigued and had their own names for these orchids.

In the past much difficulty has been encountered by naming of the same species by separate authors who saw perhaps a totally different vision. The preservation of flowers, one of the direct means of identification and cross-matching, is made very difficult because of the substance in the flower parts. Some had as many as ten synonyms.

The genus as a whole is of Central and South American origin. It is epiphytic, almost entirely so. The plants resemble other orchids from the same region and there are affiliations between them. Some, indeed, have similarities in flower organisation and appearance, among them the gongoras. One cannot do anything but wonder why such floral diversification became necessary.

In cultivation all the genus need warmth, although some appear to have sufficient stamina to withstand intermediate to cool conditions. For flowering minimum temperatures of 12 degrees (54°F.) are needed during the cooler months of the year. Plants which are adequately shaded under 60 to 75 per cent material will withstand the hottest summers, provided they are watered and not allowed to shrivel.

The plants are impossible for pots or other containers which stop the downward growth of flower spikes toward the base of the plants and further. A wire basket with a low-penetration lining is ideal. Soft layers of coconut fibre or similar material give the spikes a chance to thrust the flowering head through. The flowers, compared with the size of the pseudo-bulbs, appear out of proportion. Nearly all have poor lasting qualities, and some have a perfume that deteriorates.

Stanhopeas generally are impossible to bench and the containers must be suspended. During the annual cycle they have a dormant period, but it should not be of long duration. In this period the principal thing to watch is that they are not overwatered. Indeed, regardless of the ribbiness of pseudo-bulbs, it is better to err on the side of deprivation rather than watering regularly before new growth is noted.

The species have varying flowering periods, some of them summer, others later in the year. Appearance of flower spikes is unpredictable, their showing through the base of containers the only indicator. A light application of liquid fertiliser with every watering suits the genus and filling material for growing baskets should be based on coarse bark.

Stelis (Swartz) 1799

An uncommon genus in cultivation, although as will be noted from the generic title, it has been long known if not grown. Considering the number of species, more than four hundred, it is unusual that they have attracted so little attention. The habitats are spread over the Central American, Caribbean and South American region, nearly all the species originating in moist forest. A similarity to many pleurothallis and masdevallias is notable in their morphology.

The plants resemble bundles of leaves, in most instances the junction between leaf and petiole generating the flower raceme. The flowers are unlike most genera, being markedly three-pointed; in this they are similar to masdevallias. The three sepals outshine the other and perhaps more vital parts of flowers.

Most plants are small; the flower stems are slender, perhaps better described as wiry, and the flowers are borne over almost their total length. Despite their smallness, the flowers if seen through a telephoto lens or a magnifying glass are most beautiful. Colours vary from silvery brilliance to purple softness. Most are notable for the fine coating of hair-like cilia overlaying the sepals.

The genus could, like masdevallias, be considered moist-growing and accommodation arranged to give the plants the benefit of shade and warmth. Temperatures of 15 to 18 degrees (60–65°F.) are recommended, with humidity fairly high, about 65 to 80 per cent. In warmer climates these plants suffer if temperatures rise above 28 degrees (82°F.), and some system of cooling should be provided.

Glasshouse-grown plants are never so much a problem as those grown in open or semi-open conditions, as climates can be modified, either automatically or manually, to fit such genera as stelis. These orchids could scarcely be recommended for sub-tropical or tropical climates.

Pots to accommodate stelis may be either plastic or clay, with a preference for the latter. The potting material should be designed to suit moist-growing species, based principally on bark mixes and with charcoal or coarse gravel for opening the mixture and keeping it free draining. If grown in clay pots an advantage would be gained by double potting (see page 81), which insulates as well as conserves moisture. Nearly all these plants originate on tree branches and are open-root epiphytes, though they may be grown in pots in artificial conditions.

The pots should be placed on solid bench-tops, with constant humidity but never stagnant wetness. The plants also should be kept moist but never wet. If grown and flowered, pollination of any species, preferably not cross-pollinations, would be a very good project and a painstaking operation.

Stenocoryne (Lindley) 1843

A genus related to bifrenarias by some authors. The various synonyms and unacceptable species gathered under this generic title display the awful mess into which nomenclature descended at times. It can be attributed in great measure to the indiscriminate gathering of species by different nationalities and lack of communication, and to professional jealousies which appeared to motivate botantists. This is exemplified in Reichenbach's action in locking up his herbarium material after his death.

Stenocorynes are principally Brazilian orchids and for some obscure reason were known as bifrenarias. There is remote resemblance to that genus, but the total character of the flowers is different even if there is morphological linkage.

The genus was covered by Hoehne in his *Flora Brasilica*, but unfortunately his writings have not been widely translated. His listing of synonymy is complete and he illustrates some of the flowers in colour. They do not resemble bifrenarias and have some affinity with the shape and form of some encyclias.

Most orchids from Brazil in the early years of collecting were sent to Britain by various people from the Rio de Janeiro region. It was the advent of these plants which led Paxton and other innovators to design living conditions for them which were in contrast to those designed for the orchids from the 'East Indies'. He introduced fresh air to the plants and the results were immediate: losses were not eradicated but were at least ameliorated.

Stenocorynes were among those early plants, but with emphasis, as always, on the bigger and brighter genera they never assumed much of a role in the glasshouses of the fanciers.

Bifrenaria (Stenocoryne) vitellina (Lindley) Kraenzlin, in the late nineteenth century was incorporated in a group of orchids exhibited by W. L. Lewis and Co. which gained a silver R.H.S. Banksian Medal. The plant itself was awarded a Botanical Certificate.

The *Kew Bulletin* of 1894 also recorded *Bifrenaria (Stenocoryne) charlesworthii* (Rolfe). It was imported from Minas Geraes and flowered in the previous year. But as a genus they remained comparatively few through the history of orchid cultivation. Most of the species are intermediate to warm-growing; they flower in spring to autumn, which is a fairly long period. The acclimatisation and flowering of plants leads to strange reversals of seasons, and plants appear to take their time recovering from the movement.

The genus generally are pot plants, the potting medium bark based. Note should be made of growth and flowering, after which they should be dormant.

Symphyglossum (Schlechter) 1918

A genus having only one species and that one often recognised as a synonym for *Cochlioda sanguinea* (Bentham). *Symphyglossum sanguineum* (Reichenbach f.) has never appeared in *Sander's List of Orchid Hybrids* except under the synonym of *Cochlioda sanguinea*. For clarity on this point the *Handbook on Orchid Nomenclature and Hybridisation* should be consulted. Its correct botanical name is symphyglossum. This species is native to Peru, together with the five or six cochliodas.

Peruvian orchids are uncommon and related in large degree to those of the neighbouring state Colombia. They occur on the Andes range of mountains, particularly on the eastern slopes. There are few roads through the state, the main ones stretching north and south, with few intersections. However, the country has been more or less totally explored and most of the orchids are known and described, with some genera such as masdevallias and other botanical curiosities still being collected. There is conflict in the names of many species owing to collections by different nationalities. Transfers in taxonomy are still made fairly frequently. The genus symphyglossum is also recorded as Ecuadoran, but this is not generally credited in most literature.

Veitch lists *S. sanguineum* as an Ecuadoran discovery, but plants apparently did not reach Europe or Britain until after 1850. This could hardly be surprising, as it is one of the smallest of the genus cochlioda, as Veitch lists it. In some ways this species, like other small-flowered plants, did not suffer the depredations of collectors.

Cochliodas and the symphyglossum are listed as growing in cool to intermediate conditions; of the two the intermediate range would be more appropriate, with minimums never falling below 12 degrees (54°F.). The upper range is also critical, and efforts should be made to keep the maximum below 26 degrees (80°F.).

Potting materials for cochliodas should be bark, mixed with fairly coarse gravel to provide drainage. Mixes or combinations for the related group, including cochliodas and odontoglossums, should have a certain amount of dry leaf, preferably shredded, worked into the potting material; but do not add so much that it will clog drainage or turn the mix soggy or wet as it decays.

Shade should be of about 75 per cent and the plants preferably kept slightly moist at all times, with humidity rarely below 50 per cent.

Thunia (Reichenbach f.) 1852

A terrestrial genus of about four or five members, native to the Burmese and Indian regions, particularly the more tropical sections. Bartle Grant, although he may be a long time out of date and his naming superseded, contributed much to the knowledge of Burmese species and he considered that a few belonged to other taxa or could be classed as sub-species or varieties.

Thunias are bamboo-like, more so than other species or genera nominated as bearing this resemblance. However, the stems are quite soft; they have little substance and may need staking to hold them erect as they grow.

A primary aim should be to have the plants begin new growth and sustain it continuously in as short a time as possible to flowering stage. The beginning of growth should be in late spring and flowers should open in mid-summer to late summer. Once having flowered, the bamboo-like growths usually cast all the leaves and enter a dormant period over what is classed as winter.

After new growth occurs at the base of flowered stems, the plants should be repotted as soon as roots are noticed issuing from the bases of the new growths. Having flowered, the older canes or stems do not flower again. Once new growth has become self-supporting, these old growths slowly dehydrate and shrivel away to brown residual stumps. To keep the plants looking tidy they may then be removed.

Fast progress is the secret of growing the two commoner species, *Thunia marshalliana* (Reichenbach f.) and *T. bensoniae* (Reichenbach f.). The completed stems should be about 60 to 70 centimetres (24 to 28 inches) tall, stout and well-leafed up the length of the stem. These leaves are soft and easily burned by too bright light. About 60 to 75 per cent shade cloth should give them protection, but in bright climates may need increasing during warmer days. The plants should not be too heavily shaded.

As stems mature the flowers emerge from the apex and open consecutively, with two or three open at once. There should be six or eight buds in the sheath and flowers do not last very long, as they open in the hottest part of the year.

These orchids are subject to attack from red spider in similar fashion to calanthes, which start at much the same time. Humidity in growing either genus is important and should be of 75 per cent and more. The genus should be classed as sub-tropical, not tropical.

Potting material for thunias should be well drained and rich in humus; if sufficiently so, there is little need for nutrient watering until the flowers emerge.

Trichoglottis (Blume) 1825

An Indo-Asian genus of about thirty species, the habitats ranging from Burma through to Niugini, which has several species. The greatest aggregation and perhaps the most beautiful occurs in the Philippine islands, where twenty species grow, with many also in the Javanese-Sumatran group. It could be inferred from this that most, if not all, are orchids for tropical climates and warm enclosures. They are principally epiphytic.

The genus is monopodial, many species growing quite tall, the leaves distichous and most thickly rigid but short. The flowers emerge in singles from the main stems, principally from the upper parts. Although Blume originally designated some fifty or more species, many were later transferred to other genera and the present total appears to be between thirty and forty.

Most of the genus grow at low altitudes. At one period during the 1939–45 war I believed I found one growing on the island of Bougainville but never saw the plant in flower. This was at almost sea level and in a climate over the twelve months which was soft and warm and not excessively humid.

The genus is related to vandopsis and at times some of the species were referred to that genus. In looking at the phalaenopsis-like foliage and habit of *Vandopsis parishii* (Veitch and Reichenbach f.) (see page 214) no likeness is apparent, but other vandopsis do bear this resemblance.

Cultivation of nearly all trichoglottis is much the same as for terete vandas, epidendrums and many other genera. The stems as they elongate need some support; material like treefern will give the roots something to enter to find moisture and fertilisers. The root system is usually short and not as extensive or invasive as that of vandas.

Light requirements are high: it is impossible to give a percentage of shade cloth to suit all environments and climates, but it should not be denser than about 50 to 60 per cent. If the fleshy leaves begin to burn, shade must be increased as much as is necessary to protect the plants. You may think that these elongating stems are protected by distance from the roof, whether of shade cloth or solid coloured plastic; but they climb fast, and the danger zone extends to almost 1 metre (3 feet) below the roof.

With tropical orchids like trichoglottis, watering should be free during the growth period and the humidity should be high, say 70 per cent or more. The plants should not dry out for longer than a day or two at the root system. Temperatures should preferably never fall below about 18 degrees (65°F.), although at 15 degrees (60°F.) they will survive.

Trichopilia (Lindley) 1836

Not a numerous genus, the members are spread over a wide area of Central and South America. Generally the species are not well known or grown, although they have been cultivated plants with a history in collections of the large establishments of Britain and Europe in the nineteenth century. In general they are epiphytes, some originating at heights of 1500 to 2500 metres (5000 to 8000 feet) and growing on trees anything from 6 to 12 metres (20 to 40 feet) from the ground.

The species *T. coccinea* (Warscewicz), once known as a species in its own right, is now listed as a variety of *T. marginata* (Henfrey), together with other variants, of which there are a number. Warscewicz found this orchid in Central America in 1849. It has markedly compressed pseudo-bulbs and long leaves with a faint silvery sheen similar to the leaves of *Encyclia citrina* (La Llave and Lexarza) Dressler, and a few other orchids.

The genus trichopilia varies considerably from species to species, and some have little or no pseudo-bulbs. There is a toughened and extended rhizome system with the flowers emanating from it. Those of the bulbed forms flower from the base of the bulbs.

While many species originate in light forest at medium altitudes and are occasionally classed as cool-growing orchids, the view is mistaken: better results with regular growth and flowering patterns will be had from plants which are grown in intermediate conditions. The environment should be lightly humid and warm, with a minimum of about 15 degrees (60°F.) throughout the coldest periods of the year.

In growing orchids like trichopilias, where the plants come from varying climates, it is best to base artificial culture on the warmest-growing species rather than on a cooler-climate species which will tolerate slightly lower temperatures. The cooler-growing plants have flexibility, while those in the intermediate category will only tolerate upward variations. Shade should be moderate, about 60 to 75 per cent.

Few growers manage to get the number of flowers which the books say are possible. Usually flowers come in ones or twos, and anything more than that shows good culture in ideal conditions. Some species, of course, bear more flowers than others.

The pots may be either clay or plastic; they should be small rather than large. The root system is never prolific and must be cared for in combinations of bark of medium size, which should be infrequently watered when the plants are dormant. Liquid fertilisers should be applied in weak form during growth and flowering; about once a fortnight should be sufficient.

Vanda (Jones) 1795

In Paxton's *Botanical Dictionary* of 1868, the author of the genus is given as Robert Brown. The word vanda derives from a Sanskrit term used to denote the first of the genus recorded, namely *Vanda roxburghii* (R. Brown) and similar orchids such as saccolabiums. Several genera are included and Paxton goes on to describe the flowering of some of the species as

> producing from thirty to one hundred spikes each; indeed, we know from the best authority that there are single plants of *Saccolabium guttatum* growing upon trees in the Botanic Garden, Calcutta, which produce every year from fifty to one hundred spikes of flowers. In their native districts all, or the greater part of the plants referred to the genus, grow upon trees in dense forests, and consequently derive their chief support from the atmosphere; therefore their cultivation in this country is rendered very simple and easy, as the greatest part of them require nothing more than a piece of wood to attach themselves to, or an ornamental wire basket filled with moss and broken pots, and suspended from the pillars or roof of the house.

Other historic details of the period to about 1900 reveal information about the relative size of plants obtained from the forests of Indo-Asia and the Americas which makes the plants obtainable from similar sources in the closing years of the twentieth century seem very poor in comparison. If anything they resemble the pieces discarded so easily from the seemingly endless supply. More information about the vandaceous orchids is given in *Growing Orchids, Book Three.*

In combining the various genera to make a collection of species plants, a certain amount of discrimination must be shown. Although many vandas grow in a sub-tropical or tropical climate, the variation in foliage type is a deciding factor. Those vandas which have foliage shaped like knitting needles do not combine well with other shade-loving or diffused-light orchids such as *Vanda (Euanthe) sanderiana* (Reichenbach f.). It will be noted that the name is spelled differently in *Growing Orchids, Book Three*, when the letter 'i' was dropped from this and similar suffixes. Taxonomists made the decisions.

Rudolf Schlechter removed *Vanda sanderiana* (Reichenbach f.) from the genus and created the generic title *Euanthe* for this orchid alone.

Most vanda species originate in tropical climates, and would suit inclusion in groups of tropical orchids of various genera. It is impossible to be casual in selecting vandas for indoor or glasshouse cultivation. The plants are large when properly cultivated and they demand much more room than smaller genera such as paphiopedilums or cattleyas. They should be allowed enough space to realise their potential.

A collection of vanda species could never be as large as a mixed collection which included hybrids. Vandas could be included as part of a collection based on other vandaceous species. The vanda section would be small because of poor availability and it would be necessary to purchase or obtain the species over a number of years as they became available. The difficulties of establishing a collection of vandas could dismay many people in cooler climates where the stock is limited to what can be brought in from overseas and quarantined.

The species recommended, taking into account these necessities, include:

V. (*Euanthe*) *sanderiana* (Reichenbach f.), the first choice because of its tremendous variation in flower colour. It was named in 1882.
V. lamellata (Lindley) 1838.
V. merrillii (Ames and Quisumbing) 1932.
V. roeblingiana (Rolfe) 1894.
V. coerulescens (Griffith).

All these are strap-leaf vandas preferring middle order shade. Some growers disbelieve this restriction and grow them in sunny conditions — until the day they burn.

Terete-leaf vandas which will withstand almost any strength of sunlight include:

V. hookeriana (Reichenbach f.) (syn. *Papilionanthe hookeriana*).
V. teres (Roxburgh) Lindley (syn. *Papilionanthe teres*).

One of the most difficult vandas to grow, much less combine with other intermediate-climate orchids, is *Vanda coerulea* (Griffith ex Lindley). This Burmese species has been hand-pollinated and several clones and cultivars have been advertised at various times. *Vanda coerulea* itself is subject to most rigorous natural conditions but does not respond well to such treatment in artificial environments.

The method of cultivating a collection of vandaceous species can be varied from roofed enclosures to shade-cloth protection to completely outdoor culture with trellises or stakes for support; the root systems being allowed to proliferate into the underlay of material on which they are grown. When they are compared with small genera such as paphiopedilums, one gets an idea of the size needed to accommodate vandas. They can be grown in enclosures, but more planning is needed.

The temperature for enclosed vanda assemblies would need to be not less than 18 degrees (65°F.) in the most adverse outside conditions. Perhaps a slight drop to a lower temperature occasionally would do no harm, but the plan should be for the best range. The upper limits are well within what is normal in tropical or sub-tropical climates and present no problem.

Pots should be large containers of 30 centimetres (12 inches) or more. Slat baskets of similar size are also suitable. Once past the seedling stage, in which there may be one or two moves, there is no restriction on size for the small plants. If they are transferred to large containers once they develop proliferating root systems, there is more chance that the roots can be trained and contained.

Vandas occupy their containers for longer than most other epiphytes, so the potting material should be solid and large. There is no need to fill containers with large pieces of heavy material and the lower one-third of the pots or other containers may be filled with lighter chunks of polystyrene foam material. Most of this is inert and will not affect the roots. Fertilisers in tropical areas, where growth and flowering is almost endless, should be applied in weak formulations, frequently and regularly or solids and slow-release types may be used.

Vandopsis (Pfitzer) 1889

Of all the vandaceous orchids this appears less frequently cultivated, yet it is not so different in its requirements from any other member. Native to the western Pacific and across into Thailand, India and Burma, some ten or more species are known. Of these only two or three have attained even general acknowledgement from orchid growers. An unflowering plant in Bougainville hinted to me that the genus did stretch that far in its range.

The species *Vandopsis parishii*, originally believed to be a vanda and later to belong to the stauropsis, was described by Reichenbach in 1868. It was discovered in Burma by the Rev. Parish, who must have had a wonderful life outside his religious work. He was responsible for finding many orchids and a number were named after him. The foliage of *V. parishii* is more like that of phalaenopsis than vanda, and the vandopsis genus, of course, has more than a fleeting relationship with that genus.

Vandopsis lissochiloides (Gaudichaud) Pfitzer may be considered the most handsome of the species, with its red-spotted flowers and red buds and stems. It is native to the Philippines and the Moluccas and is the only species which grows in the Philippines. Its habitats extend to the Javanese islands and into Thailand.

Vandopsis gigantea (Lindley) Pfitzer, which also grows in Burma and Thailand, is probably the largest morphologically, and its flowers are proportionately larger than the other species. But anyone looking for size is sure to be disappointed with the flower's diameter of 4 to 5 centimetres (1½ to 2 inches).

There are two forms or varieties of *Vandopsis parishii*, with the darkest colour appearing in *Vandopsis parishii* var. *mariottiana*, shown on page 203. This orchid was subject to much speculation, some authorities considering it to be a separate species. The common form of *Vandopsis parishii* appears in *Growing Orchids Book Three.*

While plants of *Vandopsis parishii* are rather compact and not given to branching from the monopodial stem, *Vandopsis gigantea* is a much stouter plant and not quite equalled in this by *Vandopsis lissochiloides*. In cultivation the first mentioned is better suited by basket culture and can be left in one container for a number of years. The others, however, tend to outgrow slat baskets and may best be cultivated in large containers and in coarse material. The root systems of all are very coarse also and must be controlled.

Temperature in artificial culture is important; it should be warm, with minimums of about 18 degrees (65°F.). Shade should be similar to that for the strap-leaf vandas, about 65 to 75 per cent. Fertilisers should preferably be liquid and not slow-release tablets. The plants often show dormancy in cool conditions.

Zygopetalum (Hooker) 1827

A mostly Brazilian genus of terrestrial orchids considered by most growers among the easiest to cultivate. Some species show this is so. Considering the period over which zygopetalums have been cultivated, there is some misinformation that has stood the test of time and yet remains.

The species *Z. mackayi* (Hooker) is an example of this. Originally flowered and named in the 1820 decade, apparently little scope has since been allowed for variation of the species and everyone has expected it to be monotypic. There are basic differences and plants have been carrying various names over the history of the genus.

Z. mackayi is a cool-climate species and may be cultivated in quite low temperatures. A recommended minimum of about 10 degrees (50°F.) suits it very well and gives continuity. The flower spikes come with immature growth in autumn to early winter. This may be a handicap, because the plant must mature or at least continue growth over the colder part of the annual cycle and be ready to produce new growth again in the summer. This is the point at which the recommended minimum has its effect.

Z. crinitum has a distinctly different process because it commences growth in autumn and flowers in spring; it matures during summer and goes into semi-dormancy during winter. The two species are occasionally confused. There are other species, but these two more common ones are enough to satisfy most cultivators.

Potting material for the genus is similar to that for cymbidiums. The root systems are the same, fleshy and dense in well-grown plants. While zygopetalums should not be overpotted, it is better to give them room to grow and allow for at least three years' expansion. The brittle roots are easily broken and destroyed in repotting.

Shade for the genus as a whole should not be too dense, because the pale green leaves do not burn easily. The leaves are slightly broader than those of cymbidiums, with which zygopetalums may be mixed.

Marking of zygopetalum foliage is more frequently caused by cold than by heat. They are prone to develop black tips and large spots on leaves if the humidity is too high and the foliage remains wet in cold conditions.

Fertilisers, either blood-and-bone or hoof-and-horn, are best applied on the surface at the commencement of the growing season. Pellets of slow-release fertiliser may be used, but should be gone by dormancy, and liquid fertilisers will increase the texture and size of flowers. Water should be given only infrequently to all the genus over the colder and dormant period of the annual cycle, usually in winter. The plants do not suit sub-tropical or tropical climates.

Zygostates (Lindley) 1837

In this book we have taken a look at many strange orchids which do not figure prominently in collections. The last genus in the book is one of the least cultivated. It comprises several species, all apparently ill-understood, only occasionally mentioned in authoritative literature. As a genus it is principally Brazilian, perhaps Paraguayan and Argentinian, with the generic name used in synonymy for an ornithocephalus in Venezuela. Such is fame.

Hoehne, an authority on the Brazilian genera and species, has little beyond a single entry in his *Flora Brasilica Orchidaceas*, and not even that of *Zygostates lunata* (Lindley). Pabst and Dungs in *Orchidaceas Brasilienses* give synonymy for *Z. lunata* in the genera dactylostylis, ornithocephalus and a faint shot at epidendrum.

Most of these small species orchids need analysis for their mode of cultivation. *Z. lunata* is listed as a sub-tropical epiphyte, but taking that at face value could lead to many mistakes. Sub-tropical temperatures could range from 25 to 35 degrees (77–95°F.) at a low altitude, but at an altitude of about 1000 metres (3250 feet) temperatures could be reduced at times to as low as 16 degrees (62°F.).

It is always best to aim for slightly higher figures than the known minimums. For such an orchid as *Zygostates lunata*, the recommended minimum would be 15 degrees (60°F.). As a matter of interest, it was within the temperatures quoted that the plant was grown and flowered. Acclimatisation was quick, no setbacks occurred and it flowered twelve months after reception from quarantine.

Some would say that such orchids add nothing to the pastime, but the owner of this plant was more than satisfied with the price paid, which today would be considered dear. What matters in selecting and buying species is understanding the requirements of each: suitable conditions must be provided or the plants will die. It is necessary to understand the various micro-climates within an enclosure in deciding whether to suspend a plant or bench it, and where to place it. Periods of dormancy must be recognised, and treatment adjusted to suit phases of growth. Monitor the need for moisture, both in the air and at the roots, of all the plants as a group.

It is easy enough to provide correct light for most of the species described in this book, but selecting the correct degree of shade for each plant is very important. It is little use expecting to grow shade-loving genera such as masdevallias with laelias which are hungry for bright light. As an example that should suffice.

The genera and species described or illustrated in this book, when considered with those appearing in the books *Growing Orchids One*, *Two*, *Three*, *Four* and *The specialist*, comprise most of the orchids commonly grown in collections. The transparencies form part of the library of the author and were photographed over more than 40 years. The taxonomy is as correct as possible considering the many changes which occur. One must deprecate these changes, however, when they appear to be simply aimed at self-gratification with a name appended to what is virtually the work of others.

Bibliography

The following books from the private library of the author were used as reference material in writing the six books of the series:

PERIODICALS

Australian Orchid Review, 1936–88.
Orchadian (journal of the Australian Native Orchid Society), 1963–88.
Orchid Review, 1893–1988.
Orchid World, vols 1–6, 1910–16 (publication discontinued).

BOOKS

Anderson, Frank J., *Illustrated Treasury of Orchids* (reproductions of old lithographs), 1979.
Associacao Orquidofila de Sao Paulo, *Native Orchids of Brazil*, 1977.
Australian Native Orchid Society, *Checklist of Australian Native Orchid Hybrids*, 1981.
Australian Orchid Foundation, *The Orchidaceae of German New Guinea* (translated from German), 1982.
Ball, John S., *Southern African Epiphytic Orchids*, 1978.
Bateman, James, *The Orchidaceae of Mexico and Guatemala*, 1837.
Bedford, R., *A Guide to Australian Native Orchids*, 1969.
Birk, Lance A., *The Paphiopedilum Growers' Manual*, 1983.
Braem, Guido J., *The Brazilian Bifoliate Cattleyas*.
Castle, Lewis, *Orchids, Their Structure, History and Culture*, 1887.
Chow Cheng, *Formosan Orchids*, 1970.
Clements, M., *Preliminary Checklist of Australian Orchidaceae*, National Botanic Gardens, Canberra, 1982.
Colombia: A Reference Guide, issued by 7th World Orchid Conference.
Cooper, Dorothy, *New Zealand Native Orchids*, 1981.
Curtis's Botanical Magazine Reproductions, ed. S. Sprunger, 1986.
[de Oca, Rafael Montes?], *Humming Birds and Orchids* (illustrated by de Oca; translated by N. P. Wright), 1963.
Dockrill, A., *Australian Indigenous Orchids*, 1969.
Dodson, C. H., and Gentry, A. H., *Flora of the Rio Palenque*, 1978.
Dressler and Pollard, *The Genus Encyclia in Mexico*, 1970.
Du Petit-Thouars, Aubert-Aubert, *Histoire Particulaire des Plantes Orchidées*, 1822.
Dunsterville, G. C. K., and Garay, L. A., *Venezuelan Orchids*, 6 vols, 1959–76.
Encyclopaedia Britannica, 1980.
Fawcett, W., and Rendle, A. G., *Orchids of Jamaica*, 1982 (reprint).
Fitzgerald, R. D., *Australian Orchids*, 2 vols, 1977 (reprint).
Fowlie, J. A., *The Brazilian Bifoliate Cattleyas*, 1977.
——, *The Genus Sophronitis*, 1987.
Gale, J. A., *Orchids of Cuba*, 1987.
Garay, Leslie A., *The Genus Phragmipedium* (Orchid Digest Reprint), 1979.
Grant, Bartle, *The Orchids of Burma*, 1895.
Halbinger, F., *Odontoglossum in Mexico and Central America* (Orquidea), 1987.
Halle, Nicolas, *Flor de la Nouvelle Caledonie et Dependences, 8: Orchidees*, 1977.
Hamer, Fritz, *The Orchids of El Salvador*, 1981.
Handbook on Orchid Nomenclature and Registration, 1985.
Hawkes, Alex, *Encyclopaedia of Cultivated Orchids*, 1965.
Hoehne, F. C., *Flora Brasilica, Vol. 12, Orchidaceas, Parts 1 to 4*.
Holttum, E., *Flora of Malaya*, 1987 (reprint).
Japan Orchid Growers' Association, *Quality Stream 'Cattleyas'. The 800 Collection*.
Jayaweera, D. M. A., *Flora of Ceylon*, 1893–1900.
Johns, John, and Molloy, Brian, *Native Orchids of New Zealand*, 1983.

Jones, David, *Native Orchids of Australia*, 1987.
Johnson's Botanical Dictionary, 1917.
Kelleher, Jo, *Intriguing Masdevallias*, 1984.
Kupper, Walter, and Linsenmaier, Walter, *Orchids* (translated by Jean W. Little), 1961.
Laseron, Charles, *The Face of Australia* (first and revised editions), 1927–49.
Lavarack, P. D., and Gray, B., *Tropical Orchids of Australia*, 1985.
Lecoufle, Marcel, and Rose, Henri, *Orchids*, 1957.
Lindley, John, *Sertum Orchidaceum*, 1838.
Mark, Fred, *Introduction to Chinese Slipper Orchids* (Orchid Digest Reprint), 1987.
Menezes, L. C., *Cattleya Labiata (Lindley)* (Orchid Digest Reprint), 1987.
Missouri Botanic Gardens, *Orchids of Panama*, 1980.
Morris, Brian, *Epiphytic Orchids of Malawi*, 1970.
Nicholls, W. H., *Orchids of Australia*, 1969.
Oakes Ames, Correll, D. S., *Orchids of Guatemala*, 1983.
Ortiz, Pedro, and others, *Orchideas Ornamentales de Colombia*, 1980.
Pabst, G., *Rupicolous Laelias* (Orchid Digest Reprint), 1984.
Pabst, G. F. J., and Dungs, F., *Orchidaceae Brasilienses*, 2 vols, 1975–7.
Paxton's Botanical Dictionary, 1868.
Pears' Cyclopaedia, 1900, 1930.
Pocock, Maynard R., *Ground Orchids of Australia*, 1972.
Pradhan, Udai C., *Indian Orchids, Guides 1 and 2*, 1976, 1978.
Quisumbing, Dr Eduardo, *Philippine Orchids*, 1981.
Rentoul, J. N., *Growing Orchids, Book One: Cymbidiums and Slippers*, 1980.
——, *Growing Orchids, Book Two: The Cattleyas and other Epiphytes*, 1982.
——, *Growing Orchids, Book Three: Vandas, Dendrobiums and Others*, 1982.
——, *Growing Orchids, Book Four: The Australasian Families*, 1985.
——, *The Specialist Orchid Grower*, 1987.
Rhodehamel, W. A., *A Masolevallia Cultural Guide*, 1988.
Rolfe, R. A., and Hurst, C. C. H., *The Orchid Stud Book*, 1909.
Sagarik, Rapee, *Beautiful Thai Orchids*, 1975.
Sander, Frederick, *Reichenbachia*, 4 vols, 1888.
Sander's List of Orchid Hybrids, 6 vols, 1945–85.
Sander's Orchid Guide, Sanders (St Albans, UK) Ltd, 1927.
Sauleda, R., and Adams, R., *Bahama Package*, reprint of papers, 1982–7.
Saunders, W. Wilson (ed.), *Refugium Botanicum* (1869–1872–1882), 1980 reprint.
Schelpe, E., *An Introduction to South African Orchids*, 1966.
Schultes, R. E., and Pease, A. S., *Generic Names of Orchids*, 1963.
Schweinfurth, Charles, *Peruvian Orchids*, 4 vols and Supplement, 1958–70.
Seidenfaden, G., *Orchid Genera in Thailand — Dendrobium Sw.*, 1985.
Smith J. J., *Orchids of Java*, 1984.
Stewart, J., and Hennessy, E.F., *Orchids of Africa*, 1981.
Stewart, Joyce, and others, *Wild Orchids of Southern Africa*, 1982.
Sweet, Herman R., *The Genus Phalaenopsis*, 1980.
Valmayor, Helen, *Orchidiana Philippiniana*, 2 vols, 1985.
van der Pijl, L., and Dodson, C. H., *Orchid Flowers, Their Pollination and Evolution*, 1966.
Veitch, James, *A Manual of Orchidaceous Plants*, 2 vols, 1887–9 (original edn).
von Drateln, Hermann, *Odontoglossum Culture*, [1960?].
Warner, R., and Williams, B. S., *The Orchid Album*, vols 2, 3, 4, 1883–5.
Waters, V. H., and Waters, C. C., *Survey of the Slipper Orchids*, 1973.
Wiard, Leon A., *An Introduction to the Orchids of Mexico*, 1987.
Williams, Brian, and others, *Orchids for Everyone*, 1980.

The proceedings of various Orchid Conferences were also referred to for useful information and data, as well as various publications of Alfred Russell Wallace.

Reference maps used were: *South America*, Kummerley and Frey; *Indonesia, the Philippines, Malaysia, Vietnam and other Asian countries* by Robinson; *Bangladesh, Sri Lanka,*

India, including Burma by Bartholomew. Outdated references to boundaries and state changes were traced through the older encyclopaedias.

In consulting these and other references many conflicting facts were noted, and many mistakes were found, including some in what are usually considered impeccable sources. These were accepted as part of the processes of recording, interpretation and not least the printing and reproduction, and not held too unkindly against the authors.

Cross references from *The Orchid Review*, edited by Robert Allen Rolfe and others into the *Gardeners' Chronicle* were made available through the courtesy of the Melbourne Herbarium.

Index

Acampe
- *multiflora* (syn. *Vanda multiflora*), 1
- *papillosa*, *101*

Acineta
- *chrysantha* (syn. *A. densa*), 2, *101*

Ada
- *allenii*, 3, *102*
- *aurantiaca*, 3, 84
- *lehmannii*, 3

Aerangis
- *citrata*, *102*
- *mystacidii*, 13
- *somalensis*, *103*

Aeranthes
- *africana*, 5
- *arachnites*, *103*, *104*

Aerides
- *crassifolium*, 7
- *crispa*, 7, *104*
- *lawrenceae*, 7
- *multiflora*, 6, 7
- *vandarum*, 6, 7, 8
- *odorata*, 7
- *quinquevulneram*, 8

Amesiella
- *philippinense*, 9, 77, *105*, 212

Ancistrochilus
- *rothschildiana*, 10
- *thompsonianus* (syn. *Pachestoma thompsonianus*, 10, *105*

Angraecum
- *eburneum*, 11, 12
- *infundibulare*, 12
- *philippinense*, 77
- *sesquipidale*, 11, 12

Anguloa
- *cliftonii*, *106*
- *clowesii*, 14, 77
- *ruckeri*, 14, *106*
- *uniflora*, 14

Anoectochilus
- *yatesae*, 93

Ansellia
- *africana* (syn. *nilotica*, *gigantea*), 16, *107*

Arachnis
- *annamensis*, 17
- *clarkei*, 17, 18, *107*
- *flos aeris*, 17
- *lowii*, 17, 18
- *rohaniana*, 17

Arpophyllum
- *alpinum*, 19
- *spicatum*, 19, *108*

Arundina
- *graminifolia* (syn. *A. bambusifolia*), 20, *108*

Ascocentrum
- *curvifolium*, 21, *108*
- *miniatum*, 21, *109*

Aspasia
- *epidendroides*, 22
- *lunata*, 22, *109*
- *pusila*, 22
- *principissa*, 22
- *variegata*, 22

Baptistonia
- *echinata*, 23, *110*

Barkeria
- *barkeriola*, 24
- *chinensis*, 24
- *dorothea*, 24
- *elegans*, 24
- *lindleyana*, 24
- *skinneri*, 24, *111*

Bifrenaria
- *aurantiaca*, 25
- *harrisoniae*, 25, *111*
- *inodora*, 25, *112*
- *longicornis*, 26
- *maguirei*, 26
- *saxicola*, 26
- *tyriananthina*, 26
- *wittigii*, 26, *112*

Brassia
- *allenii*, 84
- *angusta*, 30
- *caudata*, 30
- *gireoudiana*, 30
- *glumacea*, 3
- *keiliana*, *113*
- *longissima*, 30
- *verrucosa*, 29, *114*

Broughtonia
- *sanguinea*, 31, *114*

Brassavola
- *cucullata*, 27
- *ceboletta*, 27
- *digbyana*, 240
- *glauca*, 240
- *nodosa*, 27, *113*
- *See also* Rhyncolaelia

Bulbophyllum
- *becarii*, 32
- *ericsonii*, 32
- *globuliforme*, 32, 33, *115*
- *guttulatum*, *115*
- *medusae*, 32
- *minutissimum*, 32, 33
- *odoardii*, 33

Calanthe
- *Dominyi*, 34
- *rosea*, 34, *116*
- Veitchii, 34
- *veratrifolia* (syn. *triplicata*), 34
- *vestita*, 34, *116*

Catasetum
- *hookeri*, *117*
- *pileatum* hybrid, *117*
- sp., 36

Cattleya
- *acklandiae*, 42
- *amethystoglossa*, 42, *117*
- *aurantiaca*, *118*
- *bicolor*, 42
- *bowringiana*, 42
- *dormaniana*, 42
- *dowiana* var. *aurea*, 42, 43
- *gaskelliana*, 42, 43
- *granulosa*, 42
- *guttata*, 42, 43
- *harrisoniae* (syn. *harrisoniana*), 42
- *intermedia*, 42
- *labiata*, 39, 41, 43
- *lawrenciana*, 42
- *leopoldii*, 43
- *loddigesii*, 40, 42
- *leuddemanniana*, 42, 43
- *maxima*, 41, 42
- *mendelii*, 39, 42, 43
- *mossiae*, 42, 43
- *Ocztictepesacuxochitl*, 43
- *percivaliana*, 42, 43, *118*
- *skinneri*, 42
- *trianaei*, 42, 43
- *walkeriana*, 40, 42
- *warneri*, 42, 43, *119*
- *warscewiczii* (syn. *gigas*), 42, 43, *119*

Ceratostylis
- *rubra*, 44, *120*

Chiloschista
- *phyllorhiza*, 44
- *usneioides*, *120*

Chondrorhynca
- *aromatica*, *121*
- *discolor*, 46

Chysis
- *aurea* (syn. *bractescens*), *121*
- sp., 36

Cirrhaea
- *dependens*, 47, *122*
- *viridipurpurea*, 47

Cischweinfia
- *pusilla*, 22

Clowesia
- *rosea*, *122*

Cochlioda
- *noetzliana*, 123
- *rosea*, *123*
- *sanguinea*, 256

Cochleanthes
- *discolor* (syn. *Chondrorhynca discolor*), *122*
- sp., 46

Coelogyne
- *cristata*, *123*

Collectanea Botanica, 1

Comparettia
- *falcata*, 48
- *macroplectron*, 48, *124*

Constantia
- *australis*, 49
- *cipoensis*, 49, *124*
- rupestris, 49

Coryanthes
- *maculata*, *125*

Cryptochilus
- *sanguinea*, 50, *125*

Cuitlauzina
pendula, 51, *125*
Cymbidium
erythrostylum, 17
finlaysonianum, ix
insigne, 17
lowianum, ix
Cycnoches
pentadactylon, *126*
sp., 36
Cynorkis
ridleyi, 52, *126*
Cypripedium
calceolus, 53
japonicum, 53, *127*

Dendrobium
aggregatum, 57, *127*
arachnites, 58
amethystoglossum, *128*
bellatulum, 56, *128*
bensoniae, 56
bigibbum, 5
var. *phalaenopsis*, 57
bullenianum, *129*
capillipes, 58, *129*
crepidatum, 57, 58
chrysanthum, 56, 57
chrysocrepis, 58
chrysotoxum, 57
cruentum, 56
crystallinum, 58
delicatulum, *130*
farmeri, 57
fimbriatum, 57
finlaysonianum, 57
formosum, 55
gibsonii, 57
heterocarpum (syn. *aureum*), 55
infundibulum, 55
lituiflorum, 57
moschatum, 55
nobile
var. *nobilius*, 56
var. *virginale*, 54
parishii, 58
pierardii, 56
sanderae, 57
thyrsiflorum, 57
var. *densiflorum*, 57
trigonopus, *132*
sanderae, 57
secundum, *130*
senile, 58, *131*
subclausum, *131*
wardianum, 58
Dendrochilum
cobbianum, *132*
filiformis, 59
glumaceum, 59, *132*
Descourtilz, 237
Dinema
polybulbon, 66
Dipodium
punctatum, 71
Dipteranthus
pustulatus, *133*
Disa
cardinalis, *133*
disa collection, *133*
nervosa, 61
triloba, 61
tripetaloides, 61, *134*
uniflora, 60, *134*
Dominyi, 34
Dracula
erythrochaete, 96
Dryadella
edwallii, *135*
simula, 95
Du Petite Thouars, 11

Euanthe
sanderiana, 17, *201*, 260
Ellis, Rev. William, 11
Encyclia
baculus, *136*
citrina, 65, *136*, 259
cordigera, 66
mariae, 65, *137*
polybulbon, 65
vitellina, 64
Epidendrum
cinnabarinum, 68
difforme, *137*
flos aeris, 17
ibaguense (syn. *radicans*), 67
matthewsii, 213
medusae, *138*
Obrienianum, 64
porpax, *166*
porphyrophyllum, 213
stenopetalum, 68, *138*
Epigenium
acuminatum, 69
amplum, 69
coelogyne, *139*
lyonii, 69
muricatum, 69
Eria
coronaria, 70, *137*
flava, 70
javanica (syn. *Tania stellata*), 70, *140*
pubescens, 70
Esmeralda
clarkei, 17, *107*
Eulophia
gracilis, 71, 72
guineensis, *140*
keithii, *140*

Galeandra
baueri, 73
devoniana, 73, 74, *141*
Gomesa
theodorea, 240
Gongora
atropurpureum, 75
galeata, *141*
gratulabunda, *142*
quinquenervis, 75
Grammangis
ellisii, 77, *142*
Grammatophyllum
scriptum, 78, *143*
speciosum, 78
papuana, 78
Grobya
amherstiae, 79, *143*

Habenaria
ciliaris, 80
fimbriata, 80
rodocheila, *144*
sussanae, 80
Haemaria
discolor, 93
Haraella
odorata, 82
retrocalla, *144*
Hartwegia
kienastii, 210
genna, 210
Hellerorchis
gomezoides, 243
paniculata, 243
Hexisia
bidentata, 83, *145*
Huntleya
meleagris, 84, *145*

Jumellea
gracillipes, *146*

Laelia
alaorii, 87, *146*
anceps, 85, 86
angeriri, 88
autumnalis, 86
bahiensis, 88, *147*
bradei, 88
briegeri, 88, *147*
blumenschenii, 88
brevicaulis, 88
cattleyodes, 88
caulescens, 88
crispata, 88
crispilabia, 88
cinnabarina, 88
cinnamomea, 88
dayana, 87
endsfeldzii, 88
esalqueana, 88
flava, 88
ghillanyi, 88
glodeniana, 88
gouldiana, 88
harpophylla, 88
itambana, 88
jongheana, 88, *148*
kautskyi, 88
liliputana, 88
lobata, *148*
longipes, 88
lucasiana, 87, 88
lundii, 88
macrobulbon, 88
mantiqueirae, 88
milleri, 86, 88
milletii, 88
mixta, 88
praestans, 88
pumila, 85, 87, *149*
purpurata, 85, front cover
reginae, 88

sincorana, 85, 87, *149*
speciosa, 86, *150*
tenebrosa, 86
Lemboglossum
apterum, 91
bictoniense, 91, *150*
cervantesii, 90
cordatum, 90, *151*
maculatum, 91
rossii, 90, *151*
Leptotes
bicolor, 92, *152*
minuta, 92
tenuis, 92
Lindleyella
aurantiaca, 25
saxicola, 26
Loureiro, Joao, 6
Ludisia
discolor, 93, *152, 153*
Lycaste
deppei, 94
skinneri (syn. *virginalis*), 94, *153*

Macodes
petiola, 93
Macomaria
Veichii, 93
Macradenia
rubescens, *154*
Masdevallia
burfordiense, *154*
coccinea, vii, 95
concinna, *155*
discolor, *155*
glandulosa, *156*
huebneri, *200*
ignea (syn. *militaris*), 95, *156*
strobellii, *157*
uniflora, *157*
veitchiana, *158*
Maxillaria
lepidota, *158*
negrescens, 99, *159*
picta, 99
porphyrostele, 98
sanderiana, 98, 99, *159*
seidelii, 98, *160*
variabilis, 98, 99
Meiracyllium
gemae, 205
gemma, 205
trinasutum, 205
wendlandii, *160*, 205
Miltonia
Bleuana, 208
bluntii, *161*
spectabilis, *161*
Miltonioides
cariniferum, *162*
sp., 207
Miltoniopsis
roezlii, 208, *163*
vexillaria, 208, *164*
warscewiczii, *162*
Moorea
irrorata, 214
Mormodes
sp., 36
Mystacidium
sanderi, 12

Nageliella
angustifolia, 210
purpurea, *164*, 210
Neobenthamia
gracilis, *165*, 211
Neofinetia
falcata, *165*, 212
Neolehmannia
peperomia (syn. *porpax*), *166*, 213, 270
porpax, 213, 270
Neomoorea
irrorata, *166*, 274
Notylia
barkeri, *167*
norae, 215

Oberonia
muelleriana, *167*
palmicola, 216, *167*
Odontoglossum
bictoniense, *150*
cirrhosum, *168*
citrosmum, *125*
crispum, *168*, 217
hallii, *169*
harryanum, *169*
rossii, *157*
stellatum, *170*
trilobum, *170*
See also Cuitlauzina, Lemboglossum)
Oncidium
brunleesianum, 23
excavatum, *171*, 200
forbesii, 171
luridum, 219
ornithorrhynchum, *172*, 219
papilio, *172*, 219
pusillum, 221
sarcodes, 23
varicosum, 220
zappii, 153
Orchidaceae of German New Guinea, Schlechter, 72
Orchidiana Philippiniana, Helen Valmayor, 44
Ornithocephalus
gladiatus, 222
grandiflorus, 222
inflexus, 222
iridifolius, *173*
Osmoglossum
pulchellum, *174*

Pabstia
jugosa, *174*, 223
Palumbina
candida, *175*, 224
Panisia
demissa, 225
tricallosa, 225
uniflora, *175*, 225
Paphiopedilum
barbigerum, *176*
dianthum, *176*
hainanensis, *177*
henryanum, *177*
insigne, 227
malipoense, 178
micranthum, 178
venustum, 227
villosum, 227
wardii, *179*
Papilionanthe
hookeriana, 261
teres, 261
Peristeria
elata, *179*, 228
Phaius
tankervilliae, 73
Phalaenopsis
esmeralda, 63
lueddemanniana, *180*
mannii, *180*
mastersii, 229
schilleriana, *181*, 230
violacea, *187*
Phragmipedium
longifolium, *182*
sargentianum, *182*
schlimii, *183*
sp., 232
Phretia
pusilla, 233
sp. unknown, *183*
Plantanthera
sussanae, 80
Platyclinis
glumacea, *132*
sp., 59
See also Dendrochilum
Pleione
formosana, *184*, 234
Pleurothallis
ospinae, 235
pterophera, *184*
tribuloides, *185*
Polystachya
luteola, 236
pubescens, *185*
Ponera
kienastil, 210
Promenaea
ovatiloba, *185*
stapelioides, *126*, 237
xanthina, *186*, 237
Psygmorchis
pusilla, 221

Reichenbach, Professor, 29
Reichenbachia, Frederick Sander, 17, 39, 209
Renanthera
imschootiana, 238
monachica, *187*
Restrepia
antennifera, 239
dayana, 235
elegans, 235
hemsleyana, *187*
Rhyncolaelia
digbyana, 27, *188*, 240
glauca, 27, 240
Rhyncostylis
coelestis, *188*, 241
gigantea, *189*, 241

retusa, 241
Robiquettia
cerina, *189*
gracillipes (syn. *tierneyana*), 242
merrillii, *189*
wassellii, 242
Rolfe, Robert A., 39
Rodrigueziella
gomezoides, *190*, 240
Rossioglossum
grande, *190*, 244
insleayi, 244
williamsianum, 244
Rudolfiella
aurantiaca, 25

Saccolabium
guttatum, 260
longifolium, *191*, 241
Sander, Frederick, 17, 39
Sander's *List of Orchid Hybrids*, 67
Sarcochilus
hartmannii, 161
Sarcopodium
amplum, 70
lyonii, 70
Scuticaria
hadwenii, *191*, 245
steelei, 245
strictifolia, 245
Sigmatostylix
guatemalensis, 246
Sobralia
decora, 247
macrantha, 247
xantholeuca, *192*
Sophronitella
violacea, *192*, 248
Sophronitis
acunae, *193*, 25
brevipedunculata, *193*, 244
cernua, *194*, 249
coccinea (syn. *grandiflora*), 237, 248, 249
mantiqueirae, *194*, 249
pterocarpa, 249
pygmaea, 249
wittigiana, 249
Spathoglottis
plicata, *195*, 251
Spiranthes
navarrensis, *195*, 252
Stanhopea
occulata, *190*, 253
Stelis
papaquerensis, *196*, 254
Stenocoryne
charlesworthii, 255
secunda, *197*
vitellina, 255
Symphyglossum
sanguineum, *197*, 256

Taeniophyllum
sp., 45
Theodorea
paniculata, 243
Thunia
bensoniae, 257
marshalliana, *198*, 257
Trichoglottis
brachiata, *198*, 258
wenzelii, *199*, 258
Trichopilia
coccinea, 259
marginata, *199*, 259
Trisetella
dressleri, 95
huebneri, *200*

Valmayor, Helen, 44
Vanda
coerulea, *200*, 261
coerulescens, *201*, 261
hookerae, 261
kimballiana, *202*
lamellata, 261
lowii, 17
merrillii, 261
multiflora, 1
roeblingiana, 261
roxburghii, 260
sanderiana, 260, 261
teres, 261
watsonii, *202*
Vandopsis
gigantea, *202*, 263
lissochiloides, 263
parishii, *203*, 258, 263

Warscewiczella
wendlandii, 46, *121*

Zygopetalum
crinitum, *203*, 264
mackayi, *204*, 264
Zygostates
lunata, *204*, 265